Mortuary Science

A Sourcebook

John F. Szabo

The Scarecrow Press, Inc.
Lanham, Maryland, and Oxford

SCARECROW PRESS, INC.

Published in the United States of America
by Scarecrow Press, Inc.
A Member of the Rowman & Littlefield Publishing Group
4720 Boston Way, Lanham, Maryland 20706
www.scarecrowpress.com

12 Hid's Copse Road
Cumnor Hill, Oxford OX2 9JJ, England

First paperback edition 2002

British Library Cataloguing in Publication Information Available

The hardback edition of this book was previously cataloged by the Library of Congress as follows:

Szabo, John F., 1968-
Mortuary Science : a sourcebook / by John F. Szabo.
p. cm.
Includes index.

1. Undertakers and undertaking—Bibliography. I. Title.
Z5994.S93 1993
[RA622]
363.7'5—dc20 93-28803

™ The paper used in this publication meets the minimum requirements of American National Standard for Information Sciences—Permanence of Paper for Printed Library Materials, ANSI/NISO Z39.48-1992.
Manufactured in the United States of America.

ISBN: 0-8108-4587-3
ISBN: 978-0-8108-4587-9

To
George John Szabo
and
Jule Patricia Canterbury Szabo

ACKNOWLEDGMENTS

Professor Emeritus Kenneth Vance, University of Michigan School of Information and Library Studies, for his encouragement and assistance in this project
and
the University of Michigan Interlibrary Loan Office and the many libraries and institutions that provided material for examination.

CONTENTS

FOREWORD

The primary purpose of this work is to provide educators, researchers, funeral industry management, funeral service personnel, and laypersons with a comprehensive, annotated bibliographical resource on the subjects central to the field of mortuary science. I have included only monographic works in this bibliography; to include article citations would simply be, for the most part, an attempt at indexing funeral trade journals. That is not the purpose of this work. Most periodical literature that covers mortuary science can be found in these industry and trade journals. Mortuary science literature published outside the realm of such publications generally tends to be superficial and popular in nature.

Several components factored into my selection criteria. Age and geography were of primary importance. I chose to include only those materials prepared between the years 1900 and 1991. Since mortuary science is a discipline that has not developed technologically at the pace and rapidity of other fields, the age of research material is not as significant as it might be in an area of study where the hardcover book is outdated by the time it is printed and distributed. I have noticed in my research that mortuary science texts often refer to works 40 and 50 years old, sometimes older. I chose 1900 as a start date for two reasons: it begins the century and also marks the publication of the first truly significant written contributions to modern developments in mortuary science. Pre-1900 materials generally are not as relevant to today's funeral industry. I also chose only those materials pertinent to United States mortuary practice. You may see some citations published outside of the U.S., but their contents do focus or relate in some way to American funeral practices. I made this limitation because most material is published in the U.S., it is the broadest in scope and subject matter, and the U.S. has the most elaborate and financially successful funeral industry in the world.

The headings of each section will indicate specifically what I considered to be the subjects central to mortuary science. Generally, I tried to determine if each citation was relevant to funeral directing, restorative art, cemeteries, cremation, or funeral rites. There are a small number of titles that deal chiefly with irrelevant subjects, but include what I consider to be information critical to the theme of this work. I did not include literature on burial rites, ceremonies, and mourning from world cultures. This is an entirely separate area of study, closer to the field of anthropology and cultural geography. Exceptions to this were Effie Bendann's *Death Customs: An Analytical Study of Burial Rites* and Habenstein and Lamers' *Funeral Customs the World Over.* I considered both to be principal works in the field and each contained information I thought would be valuable to readers of this book.

Citations included in this text were gathered from several indexes, bibliographies in book form, notes and bibliographies that were included in the works cited, and the items themselves. For access reasons I did not annotate every title. Several titles are extremely obscure and exist in only one copy. For those that were annotated, I obtained the material from individuals, funeral home collections, the University of Michigan Libraries, the Ann Arbor Public Library, or University of Michigan interlibrary loan service. Most were obtained through the latter. With each annotation, I tried not to make valuative judgments on the titles themselves. I am not a student of mortuary science, nor do I want to appear as such. The substance of each annotation was derived from the text of the material, table of contents, introduction, preface, foreword, notes, appendices, and indexes (in that order). Special attention was given to the organization of each work as well as its applicability to funeral service personnel.

John F. Szabo
Director
Robinson Public Library
Robinson, Illinois

Autopsy

Adams, J. R., and D. R. Mader. *Autopsy.* London: Lloyd-Luke; 1976.

Dhonau, C. O. *Manual of Case Analysis,* 2nd ed. Cincinnati: The Embalming Book Co.; 1928.

Editors of Consumer Reports. *Funerals: Consumers' Last Rights.* Mount Vernon, NY: Consumers' Union; 1977.
Note: Subtitled *The Consumers Union Report on Conventional Funerals and Burial...and Some Alternatives, Including Cremation, Direct Cremation, Direct Burial, and Body Donation.* Also published by Pantheon Books, New York.
This is an educational text for consumers about funerals. It offers an overview of the American funeral discussing high costs, typical funerals, emotional factors in the funeral transaction, and alternatives to the conventional funeral. Topics also covered include: selecting a funeral director, death benefits, casket selection, embalming, "extras," cemeteries, vaults, cremation, body donation, memorial societies, pre-need plans, and psychological aspects of the funeral. Appendixes are pre-need laws, active Veterans Administration National Cemeteries, embalming and restorative art procedures, embalming laws, guidelines for cremation without a funeral director, uniform donor card, autopsy procedure, transplantation, and guidelines and information for survivors.

Grollman, Earl A. *Concerning Death: A Practical Guide for the Living.* Boston: Beacon Press; 1974.

Grollman's book is a guide to dealing with the facts and emotions of death. The text contains 20 individually edited sections on the subject of death and funerals--intended primarily for consumers. Pertinent topics covered are: grief, Protestant, Catholic, and Jewish rites, legal concerns, insurance, coroners, funeral directors, cemeteries, memorials (gravemarkers), cremation, organ donation and transplantation, sympathy calls, condolence letters, widows and widowers, suicide, and death education.

Mayer, Robert G. *Embalming: History, Theory, and Practice*. Norwalk, CT: Appleton and Lange; 1990.
Mayer's book is an up-to-date, comprehensive textbook on embalming and restorative art. The contributors, too numerous to name, are all prominent mortuary science educators, funeral industry leaders, and other experts from specialized areas of the field. The text covers the following subject areas: embalming fundamentals, origins and history, professional health concerns, technical orientation, agonal and pre-embalming changes after death, chemicals and their use, anatomical considerations, vessel sites and selection, body preparation prior to arterial fluid injection, distribution and diffusion of fluids, injection and drainage techniques, cavity embalming (aspiration), preparation of body after injection, age and general body considerations, autopsied bodies, organ donors, delayed embalming, discoloration, moisture and vascular considerations, and effects of drugs. Each section ends with a list of key terms and concepts for discussion and a bibliography. Includes glossary, index, and full-text versions of some suggested readings. Illustrated with drawings and black and white photographs.

McFarland, J. *Textbook of Pathology*. Philadelphia: Saunders; 1904.

Selzer, Richard. *Mortal Lessons: Notes on the Art of Surgery*. New York: Simon and Schuster; 1974.

Note: Reprinted in 1975 and 1976.
Contains one section of particular interest to funeral service personnel: "The Corpse." The author, with a tongue-in-cheek approach, makes humorous and sarcastic references to embalming, morgues, autopsies, restorative art, and burial. A summary of the text refers to it as "a rich and stimulating blend of information, reflection, and literary-medical allusion."

Weber, D. L. *Autopsy Pathology Procedure and Protocol.* Springfield, IL: Charles C. Thomas; 1973.

Bibliographies

Continental Association of Funeral and Memorial Societies. *Bibliography of Death Education*. Washington, DC: Continental Association of Funeral and Memorial Societies.
Note: The Continental Association serves as a clearinghouse for information about the nation's memorial societies.

Continental Association of Funeral and Memorial Societies. *Bibliography of Funeral Reform*. Washington, DC: Continental Association of Funeral and Memorial Societies.

Fulton, Robert L. *Death, Grief and Bereavement: A Bibliography, 1845-1975*, 4th ed. In *The Literature of Death and Dying Series.* New York: Arno Press; 1977.

Fulton, Robert L., ed. *Death, Grief and Bereavement: Bibliography II, 1975-1980.* New York: Arno Press; 1981.
Note: Also cited as *Death, Grief, and Bereavement II: A Bibliography, 1975-1980*.
In this successor text to *Death, Grief, and Bereavement: A Bibliography, 1845-1975*, Fulton lists and indexes 5,373 additional citations. He includes a bar graph indicating

numbers of articles published per year from 1900 to 1979 dealing with death from a scientific or academic perspective. Prepared in collaboration with Margaret R. Reed.

Guthman, Robert F., Jr. and Sharon K. Womack. *Death, Dying and Grief: A Bibliography*. Waco, TX: Word Services; 1978. Note: Also cited as published in Los Altos, CA.

Harmer, Ruth Mulvey. *A Consumer Bibliography on Funerals*. Washington, DC: Continental Association of Funeral and Memorial Societies; 1977.
Note: Also published in 1977 by Celo Press, Burnsville, NC.
A brief annotated bibliography of books, pamphlets and government documents about various aspects of funeral service.

Harrah, Barbara K. and David F. Harrah. *Funeral Service: A Bibliography of Literature on Its Past, Present, and Future, the Various Means of Disposition, and Memorialization.* Metuchen, NJ: Scarecrow Press; 1976.

A listing of books, articles and other materials on funeral service and related, more generalized subjects like death, dying, bereavement, grief, and mourning.

Hinson, Maude R. *Final Report on Literature Search on the Infectious Nature of Dead Bodies for the Embalming Chemical Manufacturers Association*. Cambridge, MA: Embalming Chemical Manufacturers Association; 1968.
Note: Author was a medical research librarian in Downers Grove, IL.

John Crerar Library. *A List of Books, Pamphlets and Articles on Cremation Including the Cremation Association of America Collection*. Chicago: The John Crerar Library; 1918.
This bibliography contains most of the writings on cremation in English, French, and German, particularly those of the early modern cremation movement. The

Crerar Library possesses one of the most comprehensive collections of cremation literature in the United States.

Kutscher, Austin H., Jr. and M. A. Kutscher. *A Bibliography of Books on Death, Bereavement, Loss and Grief: 1935-1968*. New York: Health Sciences Publishing Co.; 1969.
Note: Also cited as *A Bibliography of Books on Death, Bereavement, Loss and Grief Published Since 1935, 1970.*

Kutscher, Austin H., Jr. and M. A. Kutscher. *A Comprehensive Bibliography of the Thanatology Literature*. New York: MSS Information Co.; 1976.

Miller, Albert Jay, and Michael James Acri. *Death: A Bibliographical Guide*. Metuchen, NJ: Scarecrow Press; 1977.
Contains 3,848 monograph, periodical and non-book citations on death including pamphlets, letters and editorials. Areas covered include general works, education, humanities, the medical profession and nursing experiences, religion and theology, science, social sciences, and audiovisual media. Includes an appendix of 47 audiovisual sources with their addresses. Citations are indexed by author and subject in separate sections. International scope. Intended for scholars, professionals and laypersons.

National Funeral Directors Association. *Articles, Reprints, and Comments Relating to Funeral Service*. Milwaukee: National Funeral Directors Association.

Poteet, G. Howard. *Death and Dying: A Bibliography (1950-1974).* Troy, NY: Whitson Publishing Co.; 1976.
Contains 11 pages of book citations and 164 pages of article citations. Citations are divided by subject. Includes author index.

Simpson, Michael A. *Dying, Death and Grief: A Critical Bibliography*. Pittsburgh: University of Pittsburgh Press; 1987.

Simpson, Michael A., ed. *Dying, Death and Grief: A Critically Annotated Bibliography and Source Book of Thanatology and Terminal Care.* New York: Plenum Press; 1979.

Vernick, Joel C. *Selected Bibliography on Death and Dying.* Washington, DC: US Government Printing Office; 1970. Note: Produced under the direction of the United States Public Health Service, National Institutes of Health, National Institute of Child Health and Human Development.

Wass, Hannelore, ed. *Death Education II: An Annotated Resource Guide.* Washington, DC: Hemisphere Pub. Corp.; 1985. Note: Updated version of 1980 title.

Wass, Hannelore, et al. *Death Education: An Annotated Resource Guide.* Washington, DC: Hemisphere Pub. Corp.; 1980.

Watts, Tim J. *The Funeral Industry: Regulating the Disposition of the Dead.* In *Public Administration Series,* P2454. Monticello, IL: Vance Bibliographies; 1988.
The author is public services librarian at Valparaiso University School of Law Library in Valparaiso, IN. The bibliography contains 24 monograph citations and 112 article citations on funeral industry regulation; 13 of the 24 monographs cited are United States government publications. Most of the articles are from law journals or popular periodicals. A general statement on funeral industry regulation precedes the bibliography.

Burial Rites and Ceremonies

Agee, James. *A Death in the Family.* New York: McDowell Obolensky, Inc.; 1958.

Note: This is a novel.

Anders, Rebecca. *A Look at Death.* Minneapolis: Lerner; 1977.
Note: Reprinted in 1984.
A book intended to help understand death and funeral customs.

Archer, H. G. *The Burial Service: Musical Setting.* Philadelphia: General Council Public Board; 1912.

Basevi, W. H. F. *The Burial of the Dead.* London: George Routledge and Sons; 1920.
A detailed historical, cross-cultural study of burial and cremation from prehistoric times to the twentieth century.

Bendann, Effie. *Death Customs: An Analytical Study of Burial Rites.* New York: Alfred A. Knopf; 1930.
Note: Also published in 1930 by Kegan, Paul and Co., London.
An examination of the funerary practices of many nations and religious groups from early times to the date of publication. Discusses the relationship between funerary practices and the belief and thought forms of a people. The text is divided into two sections: similarities and differences (of funeral rites and ceremonies). She covers disposal of the dead, general attitudes toward the corpse, purification, life after death, taboos, mourning, women's connection with funeral rites, totemic conceptions, destruction of property, and cult of the dead. Includes a comprehensive index and glossary of terms.

Bernardin, Joseph. *Burial Services: Revised and Updated.* Wilton, CT: Morehouse-Barlow Pub.; 1980.
Note: Author also cited as Joseph B. Bernadin.

Berrill, Margaret. *Mummies, Masks, and Mourners.* New York: Dutton; 1990.

Bishop, John P., and Edmund Wilson. *The Undertaker's Garland.* New York: Haskell House; 1974.

Bowman, Leroy. *The American Funeral: A Study in Guilt, Extravagance, and Sublimity*. Washington, DC: Public Affairs Press; 1959.
Note: Introduction by Harry A. Overstreet. Also published in 1964 by Paperback Library, New York, and reprinted by Greenwood Press in 1973 and 1975.
The first of the contemporary critiques of the funeral written from the viewpoint of a social scientist who sees the funeral as an anachronism in urban society. He advocates using cremation as a means of making a funeral more economical. Discusses the differentiation between the terms "funeral director," "undertaker," and "mortician." Bowman covers group behavior at funerals, behind-the-scenes activities, family contact with the undertaker, the undertaker's role in the community, and trends in the form and function of funerals.

Buck, Peter, and Te Rangi Hiroa. *Arts and Crafts of Hawaii: Death and Burial.* In *Special Publication Series*, No. 45(13). Honolulu: Bishop Museum; 1957.
Note: Illustrated.

Budge, E. A. Wallis. *The Book of the Dead*. New Hyde Park, NY: University Books, Inc.; 1960.

Carlson, Lisa. *Caring for Your Own Dead*. Hinesburg, VT: Upper Access Publishers; 1987.
A complete guide for those who wish to handle funeral arrangements themselves. The text is divided into three parts. Part 1 discusses home funerals, cremation, embalming, burial, body and organ donation, and legal issues. Part 2 lists laws, regulations, and services in each state. Includes names and addresses of state agencies and regulatory boards, crematories, and sites for body donation. The final part is a group of appendixes with details on death certificates, preneed spending, grieving, and consumer information on Federal Trade Commission funeral regulation rules. Also includes glossary of funeral-related terms. Text comes with a sample death certificate from North Dakota.

Cavanaugh, Sally. *How to Bury Your Own Dead in Vermont.* Vermont: Vanguard Press.
This work was the basis and inspiration for Carlson's *Caring for Your Own Dead.*

Davis, Daniel L. *What to Do When Death Comes.* New York: Federation of Reform Temples.
Discusses Jewish funeral customs and burial rites.

Dickerson, Robert B. *Final Placement: A Guide to Deaths, Funerals, and Burials of Notable Americans.* Algonac, MI: Reference Publications; 1982.
Note: Keith Irvine, series editor.

Draznin, Y. *How to Prepare for Death: A Practice Guide.* New York: Hawthorn Books; 1976.
Draznin's guide, published during what she terms "a bibliographic torrent" of major proportion of books on death. She contends in the preface that death has become society's prime nonfictional fascination. Draznin's text is indeed a practical guide, covering all of the details necessary in preparing for your own death: disposing of the body, the mortuary rites, costs, wills, insurance, and estate conservation. She also includes a 10-chapter section on coping with a death in the family, detailing what to do in the cases of sudden death, accidental death, suicide, homicide, and other circumstances. The text also includes an appendix of supplementary reading notes.

Editors of Consumer Reports. *Funerals: Consumers' Last Rights.* Mount Vernon, NY: Consumers' Union; 1977.

Note: Subtitled *The Consumers' Union Report on Conventional Funerals and Burial...and Some Alternatives, Including Cremation, Direct Cremation, Direct Burial, and Body Donation.* Also published by Pantheon Books, New York.

For annotation, see page 1.

Garrison, Webb B. *Strange Facts About Death.* Nashville: Abingdon Press; 1978.

Gordon, Anne. *Death Is for the Living.* Edinburgh, Scotland: Paul Harris Publishing; 1984.
Note: Subtitled *The Strange History of Funeral Customs.* While Gordon does make continual references to Scottish funeral rites and customs, the material is nevertheless applicable to American practices. Sections include: coffins, mort bells, funeral hospitality, burial services, mortcloths, walking funerals, hearses, gravestones, mourning, apparel, executions, and body-snatchers. All focus on the superficial aspects of the funeral--as the author claims, death is for the living.

Habenstein, Robert W., and William M. Lamers. *Funeral Customs the World Over.* Milwaukee: Bulfin Printing; 1963.
Note: Reprinted in 1974. First printing in 1960.
The authors present a detailed account of mortuary practices throughout the world. Sections are Asia, the Middle East, Africa, Oceania, Europe, Latin America, Canada, and the United States. Especially comprehensive are the chapters on Native American funeral customs and burial rites. Illustrated with photographs. Covers all aspects of funeral service: embalming, burial, funeral coaches and automobiles, funeral homes, mortuary science education, etc. Also includes details on Jewish, Latter-Day Saints, and American Gypsy funeral rites.

Habenstein, Robert W., and William M. Lamers. *The History of American Funeral Directing.* New York: Omnigraphics Inc.; 1990.
Note: Reprint of the original 1955 edition. Second printing in 1956.
This is certainly the cornerstone work on American funeral directing and mortuary practice. Though it is now somewhat dated and misses some important developments in the last half of the twentieth century, it does give a detailed history and chronological outline of the profession. Habenstein divides the texts into three

parts: early mortuary behavior, the rise of American funeral directing, and the organization of modern funeral service. He discusses the roots of modern funeral practice, medical embalmers, early coffins, burial cases and caskets, funeral transportation, the development of associations and professional organizations, institutional growth, and the "panorama" of modern funeral practice. Appendixes include past presidents of the National Funeral Directors Association of the United States, National Negro Funeral Directors Association, National Selected Morticians, and Jewish Funeral Directors of America; conference accredited colleges; and a list of funeral service journals. Particularly helpful are the comprehensive table of book contents and the index. Illustrated.

Interment Association of California. *Manual of Standard Crematory-Columbarium Practices*. Los Angeles: Interment Association of California; 1941.
Note: Revision and republication by the Cremation Association of America of a part of the *Manual of Standard Interment Practices and Standard Crematory-Columbarium Practices*.
Provides recommended policies and procedures for the operation of crematoriums and columbaria.

Jackson, Percival E. *The Law of Cadavers*, 2nd ed. Englewood Cliffs, NJ: Prentice-Hall; 1950.

Jones, E. Ray. *Funeral Manual*. Cincinnati: Standard Pub.; 1991.

Kastenbaum, Robert, ed. *Death and Dying*. New York: Arno Press; 1977.
Note: This title constitutes a 40-volume set dealing with all issues surrounding the subjects of death, dying, grief, bereavement, funerals, etc.

Lamm, Maurice. *The Jewish Way in Death and Mourning*. New York: Jonathan David Pub.; 1969.

Mannes, Myra. *Last Rights*. New York: Signet/New American Library; 1975.

Martin, Edward. *Psychology of Funeral Service*, 6th ed. Grand Junction, CO: Edward A. Martin.
Note: Third edition published in 1950.
Martin begins the text with a prologue on the importance of education in general and the necessity of mortuary education to society. He discusses a variety of aspects of funeral service from historical background to modern-day practical considerations including an introduction to psychology. He also includes sections on emotion, learning and memory, adjustment to mental conflict, grief, sentiment, religion (with an encyclopedic coverage of 11 religions of the world and 20 religious concepts), funeral rituals (burial, cremation, mutilation, dismemberment, cannibalism, abandonment, and exposure), public relations, and embalming, and a chapter on "psychology in action." Includes the Funeral Service Oath, index, and glossary.

Mitford, Jessica. *The American Way of Death*. New York: Simon and Schuster; 1963.
Note: Also published in 1963 by Fawcett Publications, Greenwich, CT. Numerous reprints.
Generally considered an expose on American funeral practices, Mitford's book is a tongue-in-cheek factual analysis of the funeral industry. This book was a bestseller when it was originally published and intended to educate the public in American mortuary practices. It accomplished that goal. Mitford discusses at length funeral costs, "artifacts" or items associated with the funeral service, profitability, cremation, funeral fashion, the profession, the clergy, and attitudes of the press toward funeral service and includes a chapter entitled "New Hope for the Dead." If anyone can make mortuary science humorous and give it popular appeal, Mitford certainly can and did. Appendixes include a directory of memorial societies and related organizations, how to organize a memorial society, eye banks, and body

donation for medical science. Includes a comprehensive index and bibliography of books, pamphlets, and magazine articles.

Morgan, Ernest. *A Manual of Death Education and Simple Burial*, 7th ed. Burnsville, NC: Celo Press; 1973.
Note: Published in 1964 as *A Manual for Simple Burial*. Also cited as *Manual of Simple Burial*. The 9th edition was published in 1980. The 10th edition was published in 1984 with the title *Dealing Creatively with Death: A Manual of Death Education and Simple Burial*. The 11th revised edition, published in 1988, carried this title also.
This booklet discusses efforts at funeral reform during the 1950s and 1960s, suggesting patterns by which, through group interaction, funerals may be made simpler and less costly. Advocates cremation as a means of disposing of the dead, though not exclusively.

Myers, John. *Manual of Funeral Procedure*. Casper, WY: Prairie Publishing Co.; 1956.

Polson, Cyril J., R. P. Brittain, and T. K. Marshall. *Disposal of the Dead*. New York: Philosophical Library; 1953.
Note: Also published in 1962 by English Universities Press, London, and by Charles C. Thomas Publishers, Springfield, IL.
A comprehensive study of burial and cremation practices, focusing somewhat on those of England. Contains a thorough historical introduction to the disposal of the dead. Also includes sections on mediate disposal (death certificates, coroners, registration, etc.), cremation, burial (churchyards, cemeteries, burial grounds), funeral rites, exhumation, embalming, and funeral direction. Distinguishes mummification and embalming as modes of preservation. Treats unusual subjects such as preservation of human heads, ship-burial, and radioactive corpses.

Poovey, W. A., ed. *Planning a Christian Funeral: A Minister's Guide*. Minneapolis: Augsburg; 1978.

A book of popular funeral sermons with an introductory text on the purpose of a funeral and the facets of the funeral and burial ceremonies. Biblical text accompanies each sermon, and at the end of each the author lists the preacher, the occasion, and comments.

Rush, Alfred C. *Death and Burial in Christian Antiquity*. Washington, DC: Catholic University of America Press; 1941.

Rutherford, Richard. *Death of a Christian: The Rite of Funerals*, rev. ed. Pueblo, CO: Pueblo Pub. Co.; 1990.
Note: Part of Studies in the Reformed Rites of the Catholic Church: Vol. 7. Originally published in 1980.

Smith, Donald Kent. *Why Not Cremation*. Philadelphia: Dorrance and Co.; 1970.
Smith begins the short book with the statement, "Today, people who die and are buried can rest assured that eventually their bodies will be dug up, thrown into a common grave, or cremated." With this, he begins his argument for cremation. He includes details on burial and cremation customs and policies in a variety of countries, obtained by writing letters to world embassies.

Sourcebook on Death and Dying, 1st ed. Chicago: Marquis Professional Publications; 1982.

Tegg, William. *The Last Act: Being the Funeral Rites of Nations and Individuals*. Detroit: Gale Research; 1973.
Note: Reprint.

Types of Funeral Services and Ceremonies. New York: National Association of Colleges of Mortuary Science, Inc.; 1961.

Van Der Zee, James, Owen Dobson, and Camille Billops. *The Harlem Book of the Dead*. New York: Morgan and Morgan; 1978.
Text and poems based on a set of photographs, taken over several years, in a Harlem mortician's funeral parlor.

Focuses on African-American funeral rituals, coffins, and remembrance pictures. Illustrated.

van Gennep, Arnold. *The Rites of Passage.* Chicago: University of Chicago Press, Phoenix Books; 1960.
Note: Also published in 1960 by Routledge and Kegan Paul, London. Paperback reprint in 1977.
Discusses the social anthropology of funeral rites and ceremonies.

Wagner, Johannes, ed. *Reforming the Rites of Death.* Vol. 32 of the *Concilium Series.* Mahwah, NJ: Paulist Press; 1968.

Wilson, Sir Arnold, and Hermann Levy. *Burial Reform and Funeral Costs.* London: Oxford University Press; 1938.

Wisconsin Funeral Service: A Consumer's Guide, 3rd ed. Milwaukee: UWIM CCA.

Burial Vaults and Caskets

Carlson, Lisa. *Caring for Your Own Dead.* Hinesburg, VT: Upper Access Publishers; 1987.
A complete guide for those who wish to handle funeral arrangements themselves. The text is divided into three parts. Part 1 discusses home funerals, cremation, embalming, burial, body and organ donation, and legal issues. Part 2 lists laws, regulations, and services in each state. Includes names and addresses of state agencies and regulatory boards, crematories, and sites for body donation. The final part is a group of appendixes with details on death certificates, preneed spending, grieving, and consumer information on Federal Trade Commission funeral regulation rules. Also includes glossary of

funeral-related terms. Text comes with a sample death certificate from North Dakota.

Cavanaugh, Sally. *How to Bury Your Own Dead in Vermont.* Vermont: Vanguard Press.
This work was the basis and inspiration for Carlson's *Caring for Your Own Dead.*

Davies, M. R. R. *The Law of Burial, Cremation and Exhumation.* London: Shaw and Son; 1956.
Note: Also published by State Mutual Book in 1974.

Editors of Consumer Reports. *Funerals: Consumers' Last Rights.* Mount Vernon, NY: Consumers' Union; 1977.

Note: Subtitled *The Consumers' Union Report on Conventional Funerals and Burial...and Some Alternatives, Including Cremation, Direct Cremation, Direct Burial, and Body Donation.* Also published by Pantheon Books, New York.

For annotation, see page 1.

Facts Every Family Should Know, 3rd ed. Forest Park, IL: Wilbert, Inc.; 1967.
Note: First and second editions published in 1960 and 1964, respectively, by Wilbert W. Haase Co., Forest Park, IL. Illustrated with burial vault information.
This short booklet, published by a burial vault manufacturer, contains information for consumers on funeral customs, burial vaults, making a will, and social security benefits, and veterans' benefits, including burial in national cemeteries. Also contains a section for recording family information and final instructions regarding funeral arrangements.

Gordon, Anne. *Death Is for the Living.* Edinburgh, Scotland: Paul Harris Publishing; 1984.
Note: Subtitled *The Strange History of Funeral Customs.* While Gordon does make continual references to Scottish

funeral rites and customs, the material is nevertheless applicable to American practices. Sections include: coffins, mort bells, funeral hospitality, burial services, mortcloths, walking funerals, hearses, gravestones, mourning, apparel, executions, and body-snatchers. All focus of the superficial aspects of the funeral--as the author claims, death is for the living.

Habenstein, Robert W., and William M. Lamers. *The History of American Funeral Directing*. New York: Omnigraphics Inc.; 1990.
Note: Reprint of the original 1955 edition. Second printing in 1956.
For annotation, see page 10.

Johnson, J. and M. *Tell Me Papa: A Family Book for Children's Questions About Death and Funerals*. Council Bluffs, IA: Centering Corporation; 1978.
Details for children on hearses, caskets, graves, and vaults. Illustrated by Shari Borum.

Jones, Barbara. *Design for Death*. Indianapolis: Bobbs-Merrill Co.; 1967.
This extensively illustrated book discusses the art, fashion, and design surrounding the subjects of death and funerals. These include: the corpse, shroud, coffin, hearse, "undertaker's shop," floral tributes, the procession, cemetery, crematorium, tomb, and relics and mementos. Filled with historical references and anecdotes.

Krieger, Wilber M. *Successful Funeral Service Management*. Englewood Cliffs, NJ: Prentice-Hall; 1951.
This text is written for both funeral home management personnel and potential funeral professionals. It applies general management principles and concepts to the funeral service business. Krieger discusses how to enter the profession (license requirements, education, personal characteristics, etc.), public attitudes toward funeral service, management responsibilities, selecting a location, setting up the organization, financing, required

investments (with furniture, fixtures, and equipment checklists), working capital, attracting business through advertising and other means, merchandising, accounting, forms to use, credits and collections, letter writing, employment policies, personnel relations, and ethics. Includes an appendix of state licensing rules for embalmers and funeral directors compiled by O. J. Willoughby, publisher of *Southern Funeral Director*.

Mitford, Jessica. *The American Way of Death*. New York: Simon and Schuster; 1963.

Note: Also published in 1963 by Fawcett Publications, Greenwich, CT. Numerous reprints.

For annotation, see page 12.

Neilson, William A. W., and C. Gaylord Watkins. *Proposals for Legislative Reform Aiding the Consumer of Funeral Industry Products and Services*. Burnsville, NC: Celo Press; 1973.
A detailed study of the laws and practices of the United States and Canada relating to funeral arrangements, written for consumers.

Plowe, Mort C., and Rudolph C. Kemppainen. *Funeral Director's Financial Handbook*. Englewood Cliffs, NJ: Prentice-Hall; 1983.
Written by a Michigan funeral director and a Michigan-based business consultant, this handbook offers several tips and suggestions on financial management for funeral professionals. General subjects covered are: managing cash flow records and a procurement system, minimizing payment delays in the probate court system, using an accountant, selecting and utilizing an investment adviser, selecting a business attorney to match professional requirements, protecting assets against personal litigation, professional incorporation, avoiding tax errors, minimizing risk in planned business expansion, and cost containment for facilities management. Also details funeral home insurance and liability and retirement fund

planning. Includes index, sample funeral home purchase record, casket selection diagram, income analysis form, and sample funeral home floor plan.

Van Der Zee, James, Owen Dodson, and Camille Billops. *The Harlem Book of the Dead*. New York: Morgan and Morgan; 1978.

Text and poems based on a set of photographs, taken over several years, in a Harlem mortician's funeral parlor. Focuses on African-American funeral rituals, coffins, and remembrance pictures of the deceased. Illustrated.

Weathers, Neil F. *Dunham's Green Book: Service for the Funeral Directors of New England*, 23rd ed. Wilmot Flat, NH: Dunham Services; 1986.

Zamzow, Dale. *Build Your Own Coffin*. Sacramento, CA: Cougar Books.
Note: Illustrated.

Cemeteries

Alden, Timothy. *A Collection of American Epitaphs and Inscriptions with Occasional Notes*. New York: Arno Press; 1976.
Note: Two volumes.
A collection of epitaphs with explanatory notes, arranged by state, by city, and alphabetically within cities by persons. A resource on early American attitudes toward death.

American Blue Book of Funeral Directors. New York: Boylston Publications; 1972.
This telephone-book-style guide is the quintessential reference book for funeral directors. It lists most funeral homes in the United States and Canada as well as several foreign funeral homes. The book also includes news briefs and articles on current issues and trends in the

profession, air shipment regulation and policies for most major world carriers, names and addresses of major scheduled airlines, crematories and cremation service sites, state association executives, professional associations in other countries, a detailed guide to United States professional associations, burial in national cemeteries, veterans' burial allowance guidelines, a list of national cemeteries, addresses and telephone numbers of United States daily newspapers, state regulation on the transportation of deceased human bodies, details on shipment of remains to and from the United States, Canadian shipping regulations, colleges of funeral service education, buyer's guide, and index.

Beable, William H. *Epitaphs: Graveyard Humor and Eulogy*. New York: Thomas Y. Crowell; 1925.

Bendann, Effie. *Death Customs: An Analytical Study of Burial Rites*. New York: Alfred A. Knopf; 1930.
Note: Also published in 1930 by Kegan, Paul and Co., London.
An examination of the funerary practices of many nations and religious groups from early times to the date of publication. Discusses the relationship between funerary practices and the belief and thought forms of a people. The text is divided into two sections: similarities and differences (of funeral rites and ceremonies). She covers disposal of the dead, general attitudes toward the corpse, purification, life after death, taboos, mourning, women's connection with funeral rites, totemic conceptions, destruction of property, and cult of the dead. Includes a comprehensive index and glossary of terms.

Bernardin, Joseph. *Burial Services: Revised and Updated*. Wilton, CT: Morehouse-Barlow Pub.; 1980.
Note: Author also cited as Joseph B. Bernadin.

Better Business Bureau. *Facts Every Family Should Know About Funerals and Interments*. New York: Better Business Bureau; 1961.

Brown, Harold W. *How to Sell Cemetery Property Before Need.* Topeka, KS: Harold W. Brown; 1975.

Carlson, Lisa. *Caring for Your Own Dead.* Hinesburg, VT: Upper Access Publishers; 1987.
A complete guide for those who wish to handle funeral arrangements themselves. The text is divided into three parts. Part 1 discusses home funerals, cremation, embalming, burial, body and organ donation, and legal issues. Part 2 lists laws, regulations, and services in each state. Includes names and addresses of state agencies and regulatory boards, crematories, and sites for body donation. The final part is a group of appendixes with details on death certificates, preneed spending, grieving, and consumer information on Federal Trade Commission funeral regulation rules. Also includes glossary of funeral-related terms. Text comes with a sample death certificate from North Dakota.

Cemeteries and Gravemarkers: Voices of American Culture. Ann Arbor, MI: UMI Research Press; 1989.

Culbertson, J., and T. Randall. *Permanent New Yorkers.* Chelsea, VT: Chelsea Green; 1987.

Dincauze, Dena. *Cremation Cemeteries in Eastern Massachusetts.* Cambridge, MA: Peabody Museum; 1968.

Draznin, Y. *How to Prepare for Death: A Practice Guide.* New York: Hawthorn Books; 1976.
Draznin's guide, published during what she terms "a bibliographic torrent" of major proportion of books on death. She contends in the preface that death has become society's prime nonfictional fascination. Draznin's text is indeed a practical guide, covering all of the details necessary in preparing for your own death: disposing of the body, the mortuary rites, costs, wills, insurance, and estate conservation. She also includes a 10-chapter section on coping with a death in the family, detailing what to do in the cases of sudden death, accidental death, suicide,

homicide, and other circumstances. The text also includes an appendix of supplementary reading notes.

Duval, Francis Y., and Ivan B. Rigby. *Early American Gravestone Art in Photographs: Two Hundred Outstanding Examples.* New York: Dover; 1979.

Editors of Consumer Reports. *Funerals: Consumers' Last Rights.* Mount Vernon, NY: Consumers' Union; 1977.

Note: Subtitled *The Consumers' Union Report on Conventional Funerals and Burial...and Some Alternatives, Including Cremation, Direct Cremation, Direct Burial, and Body Donation.* Also published by Pantheon Books, New York.

For annotation, see page 1.

Facts Every Family Should Know, 3rd ed. Forest Park, IL: Wilbert, Inc.; 1967.
Note: First and second editions published in 1960 and 1964, respectively, by Wilbert W. Haase Co., Forest Park, IL. Illustrated with burial vault information.
This short booklet, published by a burial vault manufacturer, contains information for consumers on funeral customs, burial vaults, making a will, and social security benefits, and veterans' benefits, including burial in national cemeteries. Also contains a section for recording family information and final instructions regarding funeral arrangements.

Farrell, James J. *Inventing the American Way of Death, 1830-1920.* In *American Civilization Series,* Allen F. Davis. Philadelphia: Temple University Press; 1980.
Farrell contends that death is a cultural event and that societies reveal themselves in their treatment of death. Significant sections include the development of the modern cemetery, the modernization of funeral service, and the cosmological contexts of death. Farrell describes and analyzes the development of the American way of

death. It is not like Jessica Mitford's *The American Way of Death,* which focuses on the funeral industry's profit motive. This text emphasizes the complexity of cultural change.

Forest Lawn Memorial-Park Association. *Art Guide of Forest Lawn*. Los Angeles: Forest Lawn Memorial-Park Association; 1956.

Grollman, Earl A. *Concerning Death: A Practical Guide for the Living*. Boston: Beacon Press; 1974.
Grollman's book is a guide to dealing with the facts and emotions of death. The text contains 20 individually edited sections on the subject of death and funerals--intended primarily for consumers. Pertinent topics covered are: grief, Protestant, Catholic, and Jewish rites, legal concerns, insurance, coroners, funeral directors, cemeteries, memorials (gravemarkers), cremation, organ donation and transplantation, sympathy calls, condolence letters, widows and widowers, suicide, and death education.

Hendin, David. *Death as a Fact of Life*. New York: W. W. Norton; 1973.
The author, a former medical science journalist, offers a compendium of serious information on death, from both a scientific and pop culture perspective. He recommends cremation as a way of dealing with the land-use crisis. Hendin also suggests transforming cemeteries into playgrounds.

Huber, L. V., et al. *The Cemeteries*. Volume III of *New Orleans Architecture Series,* Mary Louise Christovich, ed. Gretna, LA: Pelican Pub. Co.; 1974.
Thoroughly illustrated with photographs and drawings, this text highlights New Orleans funerary architecture. It covers cemetery masonry, ironwork, preservation, and history. Includes an appendix of cemetery locations, a selected bibliography, and index. Aboveground burial detailed extensively.

Interment Association of California. *Manual of Standard Crematory-Columbarium Practices.* Los Angeles: Interment Association of California; 1941.
Note: Revision and republication by the Cremation Association of America of a part of the *Manual of Standard Interment Practices and Standard Crematory-Columbarium Practices.*
Provides recommended policies and procedures for the operation of crematoriums and columbaria.

Jackson, Kenneth T. *Silent Cities: The Evolution of the American Cemetery*. New York: Princeton Architectural Press; 1989.

Johnson, J. and M. *Tell Me Papa: A Family Book for Children's Questions About Death and Funerals*. Council Bluffs, IA: Centering Corporation; 1978.
Details for children on hearses, caskets, graves, and vaults. Illustrated by Shari Borum.

Jones, Barbara. *Design for Death.* Indianapolis: Bobbs-Merrill Co.; 1967.
This extensively illustrated book discusses the art, fashion, and design surrounding the subjects of death and funerals. These include: the corpse, shroud, coffin, hearse, "undertaker's shop," floral tributes, the procession, cemetery, crematorium, tomb, and relics and mementos. Filled with historical references and anecdotes.

Jordan, Terry G. *Texas Graveyards: A Cultural Legacy*. Austin: University of Texas Press; 1982.

Kastenbaum, Robert, ed. *Death and Dying*. New York: Arno Press; 1977.
Note: This title constitutes a 40-volume set dealing with all issues surrounding the subjects of death, dying, grief, bereavement, funerals, etc.

Kull, Andrew. *New England Cemeteries*. Brattleboro, VT: Stephen Greene Press.

A guide to 262 New England cemeteries with information on old cemetery art. Includes detailed maps.

Lindley, Kenneth. *Of Graves and Epitaphs.* London: Hutchinson; 1965.

Mitford, Jessica. *The American Way of Death.* New York: Simon and Schuster; 1963.

Note: Also published in 1963 by Fawcett Publications, Greenwich, CT. Numerous reprints.

For annotation, see page 12.

National Yellow Book of Funeral Directors and Services. Youngstown, OH: Nomis Publications, Inc.

Polson, Cyril J., R. P. Brittain, and T. K. Marshall. *Disposal of the Dead.* New York: Philosophical Library; 1953.
Note: Also published in 1962 by English Universities Press, London, and by Charles C. Thomas Publishers, Springfield, IL.
A comprehensive study of burial and cremation practices, focusing somewhat on those of England. Contains a thorough historical introduction to the disposal of the dead. Also includes sections on mediate disposal (death certificates, coroners, registration, etc.), cremation, burial (churchyards, cemeteries, burial grounds), funeral rites, exhumation, embalming, and funeral direction. Distinguishes mummification and embalming as modes of preservation. Treats unusual subjects such as preservation of human heads, ship-burial, and radioactive corpses.

The Price of Death: A Survey Method and Consumer Guide for Funerals, Cemeteries, and Grave Markers. Washington, DC: US Government Printing Office; 1975.
Note: Consumer Survey Handbook 3. A Federal Trade Commission Publication, Seattle Regional Office.

Questions You Should Ask About Cemetery Lot Promotions. New York: Association of Better Business Bureaus.

Sloane, David C. *The Last Great Necessity: Cemeteries in American History*. Baltimore: Johns Hopkins University Press; 1991.

Sourcebook on Death and Dying, 1st ed. Chicago: Marquis Professional Publications; 1982.

Stannard, David E., ed. *Death in America*. Philadelphia: University of Pennsylvania Press; 1975.
The author, assistant professor of American studies at Yale University, has written and collected essays on attitudes toward death as a dimension of American culture. The contributors are anthropologists, cultural historians, art historians, and literary scholars. Especially pertinent to this work is Stanley French's "The Cemetery as Cultural Institution."

Stranix, E. L. *The Cemetery: An Outdoor Classroom*. Philadelphia: Con-Stran Productions; 1977.
Note: A Student Workbook, Project Kare edition.

Thomas, Susan. *What to Do, Know and Expect When a Loved One Dies*. Renton, WA: S. K. Thomas; 1984.

United States. Department of Housing and Urban Development. *Cemeteries and Open Space Reservations*. Washington, DC: US Government Printing Office; 1970.

United States. Department of Housing and Urban Development. *Commemorative Parks from Abandoned Public Cemeteries*. Washington, DC: US Government Printing Office; 1971.

Walker, G. A. *Gatherings from Graveyards*. New York: Arno Press; 1930.
Note: Reprinted in 1977.

Wallis, Charles. *Stories on Stone: A Book of American Epitaphs*. New York: Oxford University Press; 1954.

Clergy and the Church

Allen, R. Earl. *Funeral Source Book*. In *Preaching Helps Series*. Grand Rapids: Baker Books; 1984.

Bachmann, C. Charles. *Ministering to the Grief-Sufferer*. Philadelphia: Fortress Press; 1967.
Focuses on pastoral care of the bereaved, but also includes sections on the ministry of a funeral.

Baerwald, Reuben C., ed. *Hope in Grief*. St. Louis: Concordia; 1966.
Offers suggestions for making the funeral a service of worship and for developing the funeral sermon. Also contains a collection of sermons and resources.

Bane, J. D., and A. H. Kutscher. *Death and the Ministry*. New York: Schocken; 1975.

Biddle, Perry H. *Abingdon Funeral Manual*. Nashville: Abingdon Press; 1976.
Note: Revised edition. Original edition published in 1976. This text, intended for Christian ministers, contains suggestions for sermons, music, and prayers at funerals. The author includes special material for use at the funeral of a child, for suicide, and for other tragic deaths. Includes detailed liturgies, information on how to develop a church policy on funerals, and instructions on how to conduct funeral services for a variety of Protestant denominations.

Blackwood, Andrew Watterson. *The Funeral: A Source Book for Ministers*. Philadelphia: Westminster Press; 1942.
Note: Also cited as Westminister Press.

Blair, Robert. *The Minister's Funeral Handbook: A Complete Guide to Professional and Compassionate Leadership*. Grand Rapids: Baker Books; 1990.

Cadenhead, Al, Jr. *The Minister's Manual for Funerals.* Nashville: Broadman; 1988.
Cadenhead, a Baptist minister, offers an extensive array of homiletical material, suggestions for pastoral care in the funeral setting, and suggested orders of service and scripture readings with a collection of appropriate poems. He also includes a collection of prayers and benedictions as well as suggestions for further readings.

Champlin, Joseph M. *Through Death to Life: Preparing to Celebrate the Funeral Mass,* rev. ed. Notre Dame, IN: Ave Maria; 1990.

Commission on Worship of the Methodist Church. *A Manual for the Funeral.* Nashville: Abingdon Press; 1962.
Contains titles of music especially suited for the organ.

Fraser, James W. *Cremation: Is It Christian?* Neptune, NJ: Loizeaux; 1965.
Note: Strongly opposes cremation from the standpoint of evangelical theology and biblical realism.

Fulton, Robert L. *The Sacred and the Secular: Attitudes of the American Public Toward Death.* Milwaukee: Bulfin Printing; 1963.

Funeral Liturgy Planning Guide. Collegeville, MN: Liturgical Press; 1984.

Gorer, Geoffrey. *Death, Grief, and Mourning.* In *The Literature of Death and Dying Series.* New York: Doubleday; 1965.
Note: Published in 1977 by Arno Press, New York.
Though Gorer's text does cover the practicalities of death, funerals, and their aftermath, it is written primarily from anthropological and psychological viewpoints. Gorer discusses bereavement extensively, with one section devoted to types of bereavement: death of father, death of child, etc. He also covers telling children about death, the afterlife, issues surrounding the clergy and the church, body disposal (the funeral), family gatherings,

gravestones, condolences, and mourning. The four appendixes are: current and recent theories of mourning and the present material, a questionnaire with statistical tables, religious beliefs and practices: 1963 and 1950 compared, and the pornography of death. Also includes index of informants quoted more than once.

Harmon, N. B. *The Pastor's Ideal Funeral Manual.* Nashville: Abingdon-Cokesbury Press; 1942.

Hutton, Samuel W. *Minister's Funeral Manual.* Grand Rapids: Baker Books; 1968.

Jackson, Edgar N. *The Christian Funeral.* New York: Channel Press; 1966.
An analysis of the religious significance of the funeral from a Christian perspective with special emphasis on the funeral mediation.

Jackson, Edgar N. *When Someone Dies.* Philadelphia: Fortress Press; 1972.

Jowett, Mary W. *A Guide to Funeral Planning.* Independence, MO: Worship Commission, Reorganized Church of Jesus Christ of Latter Day Saints and Herald Publishing House; 1985.

Kubler-Ross, Elisabeth. *Death: The Final Stage of Growth.* In *Human Development Books: A Series in Applied Behavioral Science,* Joseph and Laurie Braga, general editors, University of Miami Medical School. Englewood Cliffs, NJ: Prentice-Hall; 1975.
Note: Also published in 1974 by Spectrum Books, New York.
A psychiatrist and well-known authority on death, Kubler-Ross discusses many areas surrounding and encompassing the subject. Rites and customs of American Indians, Jews, Hindus, and Buddhists are covered. An essay entitled "Funerals: Time for Grief and Growth" by Roy and Jane Nichols is included.

Leach, W. H. *The Cokesbury Funeral Manual.* Nashville: Cokesbury Press; 1932.
Note: Also cited as *The Cokeburn Funeral Manual.*

Leach, W. H. *The Improved Funeral Manual.* Grand Rapids: Baker Book House; 1956.

The Lord Is My Shepherd-A Book of Wake Services. Notre Dame, IN: Ave Maria Press; 1971.
Contains options to the official text for the funeral liturgy for those planning the wake and funeral Mass.

Marchal, Michael. *Parish Funerals: A Guide to the Order of Christian Church.* Chicago: Liturgy Tr. Publications; 1987.

Martin, Edward. *Psychology of Funeral Service*, 6th ed. Grand Junction, CO: Edward A. Martin.
Note: Third edition published in 1950.
Martin begins the text with a prologue on the importance of education in general and the necessity of mortuary education to society. He discusses a variety of aspects of funeral service from historical background to modern-day practical considerations including an introduction to psychology. He also includes sections on emotion, learning and memory, adjustment to mental conflict, grief, sentiment, religion (with an encyclopedic coverage of 11 religions of the world and 20 religious concepts), funeral rituals (burial, cremation, mutilation, dismemberment, cannibalism, abandonment, and exposure), public relations, and embalming, and a chapter on "psychology in action." Includes the Funeral Service Oath, index, and glossary.

Memorial Society Association of Canada. *Church Comments on Funerals.* Edmonton, Alberta, Canada: Memorial Society Association of Canada.
Note: Prepared by the Toronto Memorial Society.
Contains comments on funerals by church leaders of various denominations.

Miller, Albert Jay, and Michael James Acri. *Death: A Bibliographical Guide*. Metuchen, NJ: Scarecrow Press; 1977.
Contains 3,848 monograph, periodical and non-book citations on death including pamphlets, letters and editorials. Areas covered include general works, education, humanities, the medical profession and nursing experiences, religion and theology, science, social sciences, and audiovisual media. Includes an appendix of 47 audiovisual sources with their addresses. Citations are indexed by author and subject in separate sections. International scope. Intended for scholars, professionals and laypersons.

Office of Worship for the Presbyterian Church (U.S.A.) Staff and Cumberland Presbyterian Church Staff. *The Funeral: A Service of Witness to the Resurrection*. In *Supplemental Liturgical Resource Series*, No. 4.: Westminster John Knox; 1986.

Order of Christian Funerals: General Introduction and Pastoral Notes. In *Liturgy Documentary Series*, No. 8. Washington, DC: US Catholic; 1989.

Phipps, William E. *Cremation Concerns*. Springfield, IL: Charles C. Thomas; 1989.
This book presents a balanced analysis of the issues surrounding cremation. Topics covered include: a history of cremation (ancient pyres), religious opposition, scientific influences, reasons for renewing the practice, a contemporary outlook, Christian acceptance, memorializing options, and pre-planning advantages. Includes extensive notes, a sample Cremation Planning Form, and a subject index. Illustrated.

Poovey, W. A., ed. *Planning a Christian Funeral: A Minister's Guide*. Minneapolis: Augsburg; 1978.
A book of popular funeral sermons with an introductory text on the purpose of a funeral and the facets of the funeral and burial ceremonies. Biblical text accompanies

each sermon, and at the end of each the author lists the preacher, the occasion, and comments.

Rite of Funerals. Washington, DC: United States Catholic Conference; 1971.
This is the official text for the funeral liturgy.

Rutherford, H. R. *The Order of Christian Funerals: An Invitation to Pastoral Care.* In *American Essays in Liturgy Series.* New Haven, CT: Human Relations Area Files Press, Yale University; 1990.

A Service of Death and Resurrection. Nashville: Abingdon Press; 1979.
Note: Part of Supplemental Worship Resources, No. 7. An aid for understanding the church's ministry at the time of death and for planning and conducting the funeral service.

Shipley, Roger R. *The Consumer's Guide to Death, Dying and Bereavement* . Palm Springs, CA: ETC Publications; 1982.
Note: Illustrated.

Wagner, Johannes, ed. *Reforming the Rites of Death.* Vol. 32 of the *Concilium Series.* Mahwah, NJ: Paulist Press; 1968.

Wolfelt, Alan D. *Death and Grief: A Guide for Clergy.* Milwaukee: National Funeral Directors Association; 1990.

Consumer Information

Arvio, Raymond. *The Cost of Dying and What You Can Do About It.* New York: Harper & Row; 1974.

Backman, Allan Earnshaw. *Consumers Look at Burial Practices.* St. Cloud, MN: Council on Consumer Information; 1956.
Discusses concerns and abuses relating to burial practices.

Offers suggestions for coping with high prices and high-pressure sales tactics.

Better Business Bureau. *Alerting Bereaved Families: A Special Bulletin*. New York: Better Business Bureau; 1961.

Better Business Bureau. *Facts Every Family Should Know About Funerals and Interments*. New York: Better Business Bureau; 1961.

Better Business Bureau. *The Pre-Arrangement and Pre-Financing of Funerals*. New York: Better Business Bureau; 1960.
Note: Also published in 1963 as *Facts You Should Know, Questions You Should Ask About the Pre-Arrangement and Pre-Financing of Funerals*.

Carlson, Lisa. *Caring for Your Own Dead*. Hinesburg, VT: Upper Access Publishers; 1987.
A complete guide for those who wish to handle funeral arrangements themselves. The text is divided into three parts. Part 1 discusses home funerals, cremation, embalming, burial, body and organ donation, and legal issues. Part 2 lists laws, regulations, and services in each state. Includes names and addresses of state agencies and regulatory boards, crematories, and sites for body donation. The final part is a group of appendixes with details on death certificates, preneed spending, grieving, and consumer information on Federal Trade Commission funeral regulation rules. Also includes glossary of funeral-related terms. Text comes with a sample death certificate from North Dakota.

Coleman, Reverend William L. *It's Your Funeral*. Wheaton, IL: Tyndale House; 1979.
Coleman, a Christian pastor, encourages advance planning of death arrangements and provides a general overview of some available options.

Congress, House Select Committee on Aging. *A Guide to Funeral Planning*. Washington, DC: Government Printing Office; 1984.

Davis, Daniel L. *What to Do When Death Comes*. New York: Federation of Reform Temples.
Discusses Jewish funeral customs and burial rites.

Editors of Consumer Reports. *Funerals: Consumers' Last Rights*. Mount Vernon, NY: Consumers' Union; 1977.

Note: Subtitled *The Consumers' Union Report on Conventional Funerals and Burial...and Some Alternatives, Including Cremation, Direct Cremation, Direct Burial, and Body Donation*. Also published by Pantheon Books, New York.

For annotation, see page 1.

Facts Every Family Should Know, 3rd ed. Forest Park, IL: Wilbert, Inc.; 1967.
Note: First and second editions published in 1960 and 1964, respectively, by Wilbert W. Haase Co., Forest Park, IL. Illustrated with burial vault information.
This short booklet, published by a burial vault manufacturer, contains information for consumers on funeral customs, burial vaults, making a will, and social security benefits, and veterans' benefits, including burial in national cemeteries. Also contains a section for recording family information and final instructions regarding funeral arrangements.

Federal Trade Commission Survey of Funeral Prices in the District of Columbia. Washington, DC: Federal Trade Commission, Government Printing Office; 1974.
An extended price study illustrating variations in price and sales practices. A consumer handbook.

Grollman, Earl A. *Concerning Death: A Practical Guide for the Living*. Boston: Beacon Press; 1974.

Grollman's book is a guide to dealing with the facts and emotions of death. The text contains 20 individually edited sections on the subject of death and funerals--intended primarily for consumers. Pertinent topics covered are: grief, Protestant, Catholic, and Jewish rites, legal concerns, insurance, coroners, funeral directors, cemeteries, memorials (gravemarkers), cremation, organ donation and transplantation, sympathy calls, condolence letters, widows and widowers, suicide, and death education.

Harmer, Ruth Mulvey. *A Consumer Bibliography on Funerals.* Washington, DC: Continental Association of Funeral and Memorial Societies; 1977.
Note: Also published in 1977 by Celo Press, Burnsville, NC.
A brief annotated bibliography of books, pamphlets and government documents about various aspects of funeral service.

Hughes, Theodore E., and David Klein. *A Family Guide to Estate Planning, Funeral Arrangements, and Settling an Estate After Death.* New York: Scribner; 1983.

Johannson, Francia Faust, ed. *The Last Rights: A Look at Funerals.* Mills, MD: Owings/Maryland Center for Public Broadcasting; 1975.
Note: Part of the Consumer Survival Kit.
An assortment of articles dealing with funeral planning, cost information, alternatives, and advice.

Jordahl, Edna K. *Planning and Paying for Funerals.* St. Paul, MN: Agricultural Extension Service, University of Minnesota; 1967.
Note: Revised edition. Also cited as written by Edna K. Fordahl in 1969.
This booklet discusses necessary arrangements for funerals, how to plan for them, financing strategies, laws that relate to death, and body donation.

Margolius, Sidney. *Funeral Costs and Death Benefits*. New York: Public Affairs Committee; 1967.

Matunde, Skobi. *Crossing the Great River: A Glimpse into the Funeral Rites of African-Americans*. Philadelphia: Freeland Publications; 1990.
Analyzes and discusses a variety of funeral practices of African-Americans. Contains a list of what should be done at the funeral of a loved one. Also gives instructions for preparing a will. Illustrated.

Mitford, Jessica. *The American Way of Death*. New York: Simon and Schuster; 1963.

Note: Also published in 1963 by Fawcett Publications, Greenwich, CT. Numerous reprints.

For annotation, see page 12.

Morgan, Ernest. *A Manual of Death Education and Simple Burial*, 7th ed. Burnsville, NC: Celo Press; 1973.
Note: Published in 1964 as *A Manual for Simple Burial*. Also cited as *Manual of Simple Burial*. The 9th edition was published in 1980. The 10th edition was published in 1984 with the title *Dealing Creatively with Death: A Manual of Death Education and Simple Burial*. The 11th revised edition, published in 1988, carried this title also.
This booklet discusses efforts at funeral reform during the 1950s and 1960s, suggesting patterns by which, through group interaction, funerals may be made simpler and less costly. Advocates cremation as a means of disposing of the dead, though not exclusively.

Myers, James, Jr. *Cooperative Funeral Associations*. New York: Cooperative League of the U.S.A.; 1946.
Note: Pamphlet 409. Author also cited as James Meyers, Jr. Also cited as published in Chicago.
An early publication indicating some reasons for high-priced funerals and offering suggestions on how consumers might cut costs.

National Funeral Directors Association. *When a Death Occurs: Needs...Concerns...Decisions*. Milwaukee: National Funeral Directors Association; 1974.

Nelson, Thomas. *It's Your Choice: The Practical Guide to Planning a Funeral.* Washington, DC: American Association of Retired Persons (AARP) and National Retired Teachers Association (NRTA); 1987.
Note: Also published in 1982 as *Your Choice: The Practical Guide to Planning a Funeral* by Scott Foresman and Co., Glenview, IL. The 1982 title is also cited as *It's Your Choice: The Practical Guide to Funeral Planning.*

Norrgard, Lee E., and Jo DeMars. *Making End-of-Life Decisions*. In *Choices and Challenges Series.* New York: ABC-CLIO; 1991.

NRTA-AARP. *A Consumer's Guide to Funeral Planning* . Washington, DC: NRTA-AARP; 1980.

The Price of Death: A Survey Method and Consumer Guide for Funerals, Cemeteries, and Grave Markers. Washington, DC: US Government Printing Office; 1975.
Note: Consumer Survey Handbook 3. A Federal Trade Commission Publication, Seattle Regional Office.

Questions You Should Ask About Cemetery Lot Promotions. New York: Association of Better Business Bureaus.

Riley, Miles O'Brien. *Set Your House in Order.* New York: Doubleday; 1980.

Simpson, Michael A. *The Facts of Death.* Englewood Cliffs, NJ: Spectrum/Prentice-Hall; 1979.
Offers information on how to plan one's funeral and estate and how to avoid unscrupulous funeral directors.

Smith, Curtis A. *Help for the Bereaved: What the Family Should Know.* Chicago: Adams Press; 1972.
Discusses funerals, death certificates, financial affairs, and benefits.

Sourcebook on Death and Dying, 1st ed. Chicago: Marquis Professional Publications; 1982.

University of Wisconsin. *Wisconsin Funeral Service: A Consumer's Guide*, 3rd ed. Madison, WI: University of Wisconsin; 1987.

Wesner, Maralene and Miles. *A Time to Weep: Funeral and Grief Messages*. Idabel, OK: Diversity OKLA; 1988.

Wolfelt, Alan D. *Death and Grief: A Guide for Clergy*. Milwaukee: National Funeral Directors Association; 1990.

Cremation

American Blue Book of Funeral Directors. New York: Boylston Publications; 1972.

For annotation, see page 20.

Basevi, W. H. F. *The Burial of the Dead*. London: George Routledge and Sons; 1920.
A detailed historical, cross-cultural study of burial and cremation from prehistoric times to the twentieth century.

Bowman, Leroy. *The American Funeral: A Study in Guilt, Extravagance, and Sublimity*. Washington, DC: Public Affairs Press; 1959.
Note: Introduction by Harry A. Overstreet. Also published in 1964 by Paperback Library, New York, and reprinted by Greenwood Press in 1973 and 1975.
The first of the contemporary critiques of the funeral written from the viewpoint of a social scientist who sees the funeral as an anachronism in urban society. He advocates using cremation as a means of making a funeral more economical. Discusses the differentiation between the terms "funeral director," "undertaker," and "mortician." Bowman covers group behavior at funerals,

behind-the-scenes activities, family contact with the undertaker, the undertaker's role in the community, and trends in the form and function of funerals.

Carlson, Lisa. *Caring for Your Own Dead.* Hinesburg, VT: Upper Access Publishers; 1987.
A complete guide for those who wish to handle funeral arrangements themselves. The text is divided into three parts. Part 1 discusses home funerals, cremation, embalming, burial, body and organ donation, and legal issues. Part 2 lists laws, regulations, and services in each state. Includes names and addresses of state agencies and regulatory boards, crematories, and sites for body donation. The final part is a group of appendixes with details on death certificates, preneed spending, grieving, and consumer information on Federal Trade Commission funeral regulation rules. Also includes glossary of funeral-related terms. Text comes with a sample death certificate from North Dakota.

Cobb, John Storer. *A Quarter-Century of Cremation in North America.* Boston: Knight and Millet; 1901.
A historical review of the foundation of cremation societies and crematoriums. Gives details on reasons behind the development of cremation.

Davies, M. R. R. *The Law of Burial, Cremation and Exhumation.* London: Shaw and Son; 1956.
Note: Also published by State Mutual Book in 1974.

Dincauze, Dena. *Cremation Cemeteries in Eastern Massachusetts.* Cambridge, MA: Peabody Museum; 1968.

Dowd, Quincy L. *Funeral Management and Costs.* Chicago: University of Chicago Press; 1921.
Note: Subtitled *A World Survey of Burial and Cremation.*
An early social scientific study of the funeral. Discusses modern development of cremation and witnessing a cremation.

Editors of Consumer Reports. *Funerals: Consumers' Last Rights.* Mount Vernon, NY: Consumers' Union; 1977.

Note: Subtitled *The Consumers' Union Report on Conventional Funerals and Burial...and Some Alternatives, Including Cremation, Direct Cremation, Direct Burial, and Body Donation.* Also published by Pantheon Books, New York.

For annotation, see page 1.

Fraser, James W. *Cremation: Is It Christian?* Neptune, NJ: Loizeaux; 1965.
Note: Strongly opposes cremation from the standpoint of evangelical theology and biblical realism.

Habenstein, Robert W., and William M. Lamers. *The History of American Funeral Directing.* New York: Omnigraphics Inc.; 1990.
Note: Reprint of the original 1955 edition. Second printing in 1956.
For annotation, see page 10.

Habenstein, Robert W. *A Sociological Study of the Cremation Movement in America* [unpublished M.A. thesis, University of Chicago, Department of Sociology, 1949] ; 1949.

Hendin, David. *Death as a Fact of Life.* New York: W. W. Norton; 1973.
The author, a former medical science journalist, offers a compendium of serious information on death, from both a scientific and pop culture perspective. He recommends cremation as a way of dealing with the land-use crisis. Hendin also suggests transforming cemeteries into playgrounds.

Interment Association of California. *Manual of Standard Crematory-Columbarium Practices.* Los Angeles: Interment Association of California; 1941.
Note: Revision and republication by the Cremation

Association of America of a part of the *Manual of Standard Interment Practices and Standard Crematory-Columbarium Practices.*
Provides recommended policies and procedures for the operation of crematoriums and columbaria.

Irion, Paul E. *Cremation*. Philadelphia: Fortress Press; 1968.

Irion, Paul E. *The Funeral: Vestige or Value?* In *The Literature of Death and Dying Series.* New York: Arno Press; 1966.
Note: Also published with Abingdon Press, Nashville, in the same year. Reprinted by Arno Press in 1976.
Based on religious, cultural, social, and psychological understanding of the nature of the funeral. Contemporary practices are evaluated in light of the valuable functions of the funeral, and new designs are proposed to conserve significant values. Contains sections on cremation and memorial societies.

John Crerar Library. *A List of Books, Pamphlets and Articles on Cremation Including the Cremation Association of America Collection*. Chicago: The John Crerar Library; 1918.
This bibliography contains most of the writings on cremation in English, French, and German, particularly those of the early modern cremation movement. The Crerar Library possesses one of the most comprehensive collections of cremation literature in the United States.

Kastenbaum, Robert, ed. *Death and Dying.* New York: Arno Press; 1977.
Note: This title constitutes a 40-volume set dealing with all issues surrounding the subjects of death, dying, grief, bereavement, funerals, etc.

Kubasak, Mike. *Cremation and the Funeral Director.* Milwaukee: National Funeral Directors Association.

Market for Funeral and Cremation Services. New York: Business Trends; 1985.

Michigan Cremation Association, Detroit. *Detroit Crematorium and Columbarium*. Detroit: Michigan Cremation Association; 1912.

Mitford, Jessica. *The American Way of Death*. New York: Simon and Schuster; 1963.

Note: Also published in 1963 by Fawcett Publications, Greenwich, CT. Numerous reprints.

For annotation, see page 12.

Morgan, Ernest. *A Manual of Death Education and Simple Burial*, 7th ed. Burnsville, NC: Celo Press; 1973.
Note: Published in 1964 as *A Manual for Simple Burial*. Also cited as *Manual of Simple Burial*. The 9th edition was published in 1980. The 10th edition was published in 1984 with the title *Dealing Creatively with Death: A Manual of Death Education and Simple Burial*. The 11th revised edition, published in 1988, carried this title also.
This booklet discusses efforts at funeral reform during the 1950s and 1960s, suggesting patterns by which, through group interaction, funerals may be made simpler and less costly. Advocates cremation as a means of disposing of the dead, though not exclusively.

National Yellow Book of Funeral Directors and Services. Youngstown, OH: Nomis Publications, Inc.

Phipps, William E. *Cremation Concerns*. Springfield, IL: Charles C. Thomas; 1989.
This book presents a balanced analysis of the issues surrounding cremation. Topics covered include: a history of cremation (ancient pyres), religious opposition, scientific influences, reasons for renewing the practice, a contemporary outlook, Christian acceptance, memorializing options, and pre-planning advantages. Includes extensive notes, a sample Cremation Planning Form, and a subject index. Illustrated.

Polson, Cyril J., R. P. Brittain, and T. K. Marshall. *Disposal of the Dead.* New York: Philosophical Library; 1953.
Note: Also published in 1962 by English Universities Press, London, and by Charles C. Thomas Publishers, Springfield, IL.
A comprehensive study of burial and cremation practices, focusing somewhat on those of England. Contains a thorough historical introduction to the disposal of the dead. Also includes sections on mediate disposal (death certificates, coroners, registration, etc.), cremation, burial (churchyards, cemeteries, burial grounds), funeral rites, exhumation, embalming, and funeral direction. Distinguishes mummification and embalming as modes of preservation. Treats unusual subjects such as preservation of human heads, ship-burial, and radioactive corpses.

Smith, Donald Kent. *Why Not Cremation.* Philadelphia: Dorrance and Co.; 1970.
Smith begins the short book with the statement, "Today, people who die and are buried can rest assured that eventually their bodies will be dug up, thrown into a common grave, or cremated." With this, he begins his argument for cremation. He includes details on burial and cremation customs and policies in a variety of countries, obtained by writing letters to world embassies.

Sourcebook on Death and Dying, 1st ed. Chicago: Marquis Professional Publications; 1982.

United States. Department of Commerce. Bureau of the Census. *1982 Census of Service Industries. Preliminary Report. Industry Series. Funeral Services and Crematories.* Washington, DC: Department of Commerce, Bureau of the Census; 1984.

Death Education

Agee, James. *A Death in the Family.* New York: McDowell Obolensky, Inc.; 1958.

Note: This is a novel.

Anders, Rebecca. *A Look at Death.* Minneapolis: Lerner; 1977.
Note: Reprinted in 1984.
A book intended to help individuals understand death and funeral customs.

Armstrong, H. G. *The American Way of Dying.* Hicksville, NY: Exposition Press; 1978.

Bayley, Joseph. *The View from the Hearse.* Elgin, IL: David D. Cooke; 1969.

Bloch, Maurice, and Jonathan Parry, eds. *Death and the Regeneration of Life.* New York: Cambridge University Press; 1982.

Clemens, Christopher, and Mark Smith. *Death: Grim Realities and Comic Relief.* New York: Delacorte; 1982.

Continental Association of Funeral and Memorial Societies. *Bibliography of Death Education.* Washington, DC: Continental Association of Funeral and Memorial Societies.
Note: The Continental Association serves as a clearinghouse for information about the nation's memorial societies.

Dempsey, D. *The Way We Die: An Investigation of Death and Dying in America Today.* New York: McGraw-Hill; 1975.

Engram, Sara. *Mortal Matters: When a Loved One Dies.* Kansas City, MO: Andrews and McMeel; 1990.

Farrell, James J. *Inventing the American Way of Death, 1830-1920.* In *American Civilization Series,* Allen F. Davis. Philadelphia: Temple University Press; 1980.

Farrell contends that death is a cultural event and that societies reveal themselves in their treatment of death. Significant sections include the development of the modern cemetery, the modernization of funeral service, and the cosmological contexts of death. Farrell describes and analyzes the development of the American way of death. It is not like Jessica Mitford's *The American Way of Death*,which focuses on the funeral industry's profit motive. This text emphasizes the complexity of cultural change.

Feifel, Herman. *The Meaning of Death*. New York: McGraw-Hill; 1959.

Feifel, Herman. *New Meanings of Death*. New York: McGraw-Hill; 1977.

Fulton, Robert. *A Compilation of Studies of Attitudes Toward Death, Funerals, Funeral Directors* [privately printed]; 1967.
Note: Reprinted in 1971 by the Center for Death Education and Research, University of Minnesota Press, Minneapolis.

Fulton, Robert, and R. Bendiksen. *Death and Identity*. Bowie, MD: Charles Press; 1976.
Note: Revised edition.

Fulton, Robert L. *The Sacred and the Secular: Attitudes of the American Public Toward Death*. Milwaukee: Bulfin Printing; 1963.

Grollman, Earl A. *Concerning Death: A Practical Guide for the Living*. Boston: Beacon Press; 1974.
Grollman's book is a guide to dealing with the facts and emotions of death. The text contains 20 individually edited sections on the subject of death and funerals--intended primarily for consumers. Pertinent topics covered are: grief, Protestant, Catholic, and Jewish rites, legal concerns, insurance, coroners, funeral directors, cemeteries, memorials (gravemarkers), cremation, organ

donation and transplantation, sympathy calls, condolence letters, widows and widowers, suicide, and death education.

Halporn, R. *The Thanatology Library*. New York: Highly Specialized Promotions; 1976.

Hendin, David. *Death as a Fact of Life*. New York: W. W. Norton; 1973.
The author, a former medical science journalist, offers a compendium of serious information on death, from both a scientific and pop culture perspective. He recommends cremation as a way of dealing with the land-use crisis. Hendin also suggests transforming cemeteries into playgrounds.

Hollingsworth, C. E., and R. O. Pasnau, eds. *The Family in Mourning: A Guide for Health Professionals*. New York: Grune and Stratton; 1977.
This book, although intended for health services personnel, contains information pertinent to funeral directors.

Irion, Paul E. *The Funeral: An Experience of Value*. Lancaster, PA: Theological Seminary; 1956.
Note: Also cited as a publication of the National Funeral Directors Association, Milwaukee.

Irion, Paul E. *The Funeral: Vestige or Value?* In *The Literature of Death and Dying Series*. New York: Arno Press; 1966.
Note: Also published with Abingdon Press, Nashville, in the same year. Reprinted by Arno Press in 1976.
Based on religious, cultural, social, and psychological understanding of the nature of the funeral. Contemporary practices are evaluated in light of the valuable functions of the funeral, and new designs are proposed to conserve significant values. Contains sections on cremation and memorial societies.

Irion, Paul E. *The Funeral and the Mourners*. Nashville: Abingdon Press; 1954.
Note: Author also cited as "Ernest F. Irion" for this title.

Jackson, C. O., ed. *Passing: The Vision of Death in America.* Westport, CT: Greenwood Press; 1977.

Johnson, J. and M. *Tell Me Papa: A Family Book for Children's Questions About Death and Funerals*. Council Bluffs, IA: Centering Corporation; 1978.
Details for children on hearses, caskets, graves, and vaults. Illustrated by Shari Borum.

Kastenbaum, Robert. *Death, Society, and Human Experience*, 3rd ed. Columbus, OH: Charles E. Merrill; 1986.
Note: Second edition published by Mosby in 1981.

Kastenbaum, Robert, ed. *Death and Dying*. New York: Arno Press; 1977.
Note: This title constitutes a 40-volume set dealing with all issues surrounding the subjects of death, dying, grief, bereavement, funerals, etc.

Kubler-Ross, Elisabeth. *Death: The Final Stage of Growth*. In *Human Development Books: A Series in Applied Behavioral Science*, Joseph and Laurie Braga, general editors, University of Miami Medical School. Englewood Cliffs, NJ: Prentice-Hall; 1975.
Note: Also published in 1974 by Spectrum Books, New York.
A psychiatrist and well-known authority on death, Kubler-Ross discusses many areas surrounding and encompassing the subject. Rites and customs of American Indians, Jews, Hindus, and Buddhists are covered. An essay entitled "Funerals: Time for Grief and Growth" by Roy and Jane Nichols is included.

Kubler-Ross, Elisabeth. *On Death and Dying*. New York: Macmillan; 1969.

Lamers, William, Jr. *Death, Grief, Mourning, the Funeral and the Child*. Chicago: National Association of Funeral Directors; 1965.

Leming, Michael R. *Understanding Dying, Death, and Bereavement*. New York: Holt, Rinehart, and Winston; 1985.

Levine, Stephen. *Who Dies*. Garden City, NY: Anchor Press-Doubleday; 1982.

Mack, A., ed. *Death in American Experience*. New York: Schocken; 1973.

Morgan, Ernest. *A Manual of Death Education and Simple Burial*, 7th ed. Burnsville, NC: Celo Press; 1973.
Note: Published in 1964 as *A Manual for Simple Burial*. Also cited as *Manual of Simple Burial*. The 9th edition was published in 1980. The 10th edition was published in 1984 with the title *Dealing Creatively with Death: A Manual of Death Education and Simple Burial*. The 11th revised edition, published in 1988, carried this title also.
This booklet discusses efforts at funeral reform during the 1950s and 1960s, suggesting patterns by which, through group interaction, funerals may be made simpler and less costly. Advocates cremation as a means of disposing of the dead, though not exclusively.

Osborne, Ernest. *When You Lose a Loved One*. New York: Public Affairs Committee; 1965.
Note: Pamphlet 269.
A discussion of the emotional, social, and financial problems relating to death in the family.

Raether, Howard C., and Robert C. Slater. *The Funeral: Facing Death as an Experience of Life*. Milwaukee: National Funeral Directors Association; 1974.

Ragon, M. *The Space of Death*. Charlottesville, VA: University Press of Virginia; 1983.
Note: Translated by Alan Sheridan.

A Service of Death and Resurrection. Nashville: Abingdon Press; 1979.
Note: Part of Supplemental Worship Resources, No. 7. An aid for understanding the church's ministry at the time of death and for planning and conducting the funeral service.

Sourcebook on Death and Dying, 1st ed. Chicago: Marquis Professional Publications; 1982.

Stannard, David E., ed. *Death in America*. Philadelphia: University of Pennsylvania Press; 1975.
The author, assistant professor of American studies at Yale University, has written and collected essays on attitudes toward death as a dimension of American culture. The contributors are anthropologists, cultural historians, art historians, and literary scholars. Especially pertinent to this work is Stanley French's "The Cemetery as Cultural Institution."

Steinfels, P., and R. M. Veatch. *Death Inside Out*. New York: Harper and Row; 1975.

Sudnow, David. *Passing On*. Englewood Cliffs, NJ: Prentice-Hall; 1967.
Note: Subtitled *The Social Organization of Dying*. Reprinted in 1969.

Warner, W. Lloyd. *The Living and the Dead*. New Haven, CT: Yale University Press; 1959.

Wass, Hannelore, ed. *Death Education II: An Annotated Resource Guide*. Washington, DC: Hemisphere Pub. Corp.; 1985.
Note: Updated version of 1980 title.

Wass, Hannelore, ed. *Dying: Facing the Facts*. New York: Hemisphere Pub. and McGraw-Hill; 1979.
Note: Includes important text on the physiology of dying by R. A. Redding. Second edition published in 1988.

Wass, Hannelore, et al. *Death Education: An Annotated Resource Guide.* Washington, DC: Hemisphere Pub.; 1980.

When Death Comes. Corvallis, OR: Oregon State University; 1963. Note: Extension Bulletin 809.

Wolfelt, Alan D. *Death and Grief: A Guide for Clergy.* Milwaukee: National Funeral Directors Association; 1990.

Disaster Planning

Federal Civil Defense Administration. *Mortuary Science in Civil Defense* (TM-11-12 [Technical Manual]). Washington, DC: Government Printing Office; 1956.

Discusses the establishment of civil defense mortuary services. Disaster plans, staffing and management concerns, training staff, identification of the dead, and interment are covered. Includes mortuary services flow chart illustrating communication and body transport patterns in a disaster setting.

Pine, Vanderlyn R. *Responding to Disaster.* Milwaukee: Bulfin Press; 1974.

Embalming

Adair, Maude Adams. *The Techniques of Restorative Art.* Dubuque, IA: W. C. Brown Co.; 1948.

American Blue Book of Funeral Directors. New York: Boylston Publications; 1972.

For annotation, see page 20.

Basevi, W. H. F. *The Burial of the Dead.* London: George Routledge and Sons; 1920.

A detailed historical, cross-cultural study of burial and cremation from prehistoric times to the twentieth century.

Bendann, Effie. *Death Customs: An Analytical Study of Burial Rites.* New York: Alfred A. Knopf; 1930.
Note: Also published in 1930 by Kegan, Paul and Co., London.
An examination of the funerary practices of many nations and religious groups from early times to the date of publication. Discusses the relationship between funerary practices and the belief and thought forms of a people. The text is divided into two sections: similarities and differences (of funeral rites and ceremonies). She covers disposal of the dead, general attitudes toward the corpse, purification, life after death, taboos, mourning, women's connection with funeral rites, totemic conceptions, destruction of property, and cult of the dead. Includes a comprehensive index and glossary of terms.

Block, S. S. *Disinfection, Sterilization, and Preservation,* 3rd ed. Philadelphia: Lea and Febiger; 1983.

Budge, E. A. Wallis. *The Book of the Dead.* New Hyde Park, NY: University Books, Inc.; 1960.

Carlson, Lisa. *Caring for Your Own Dead.* Hinesburg, VT: Upper Access Publishers; 1987.
A complete guide for those who wish to handle funeral arrangements themselves. The text is divided into three parts. Part 1 discusses home funerals, cremation, embalming, burial, body and organ donation, and legal issues. Part 2 lists laws, regulations, and services in each state. Includes names and addresses of state agencies and regulatory boards, crematories, and sites for body donation. The final part is a group of appendixes with details on death certificates, preneed spending, grieving, and consumer information on Federal Trade Commission funeral regulation rules. Also includes glossary of funeral-related terms. Text comes with a sample death certificate from North Dakota.

Cavanaugh, Sally. *How to Bury Your Own Dead in Vermont.* Vermont: Vanguard Press.
This work was the basis and inspiration for Carlson's *Caring for Your Own Dead.*

Clarke, C. H. *Practical Embalming.* Cincinnati: C. H. Clarke Publishers; 1917.

Clarke, Joseph H. *Reminiscences of Early Embalming.* New York: The Sunnyside; 1917.
An important reference work on the rise of mortuary science education in America.

Dhonau, C. O. *The ABC's of Pressure and Distribution, File 86.* Cincinnati: Cincinnati College of Embalming.

Dhonau, C. O. *Defining Embalming Fluid, File 100.* Cincinnati: Cincinnati College of Mortuary Science; 1928.

Dhonau, C. O. *Manual of Case Analysis,* 2nd ed. Cincinnati: The Embalming Book Co.; 1928.

Dhonau-Prager. *Restorative Art.* Springfield, OH: Champion Chemical Co.; 1932.

Dodge, A. Johnson. *The Practical Embalmer.* Boston: A. Johnson Dodge Publisher; 1908.

Dorn, James M., and Barbara M. Hopkins. *Thanatochemistry: A Survey of General, Organic, and Biochemistry for Funeral Service Professionals.* Reston, VA: Reston Publishing Co.; 1985.
Note: The authors are funeral service educators at the Cincinnati College of Mortuary Science.
Contains extensive sections on general chemistry, organic chemistry, and biochemistry. Discusses a variety of problems and concerns relating to the embalming process. They include: decomposition, denaturation, enzyme activity, and rigor mortis. Appendixes contain sections on radiation chemistry and a summary of the

action and composition of embalming fluids. A textbook approach with chapter summaries and questions.

Eckels College of Mortuary Science Inc. *Modern Mortuary Science*, 4th ed. Philadelphia: Westbrook Publishing Co.; 1958.

Eckels, Howard S. *Practical Embalmer*. Philadelphia: H. S. Eckels Co. Publishers; 1903.

Eckels, Howard S. *Sanitary Science: A Reference and Textbook for the Communicable Diseases, Disinfection and Chemistry for the Undertaker*. Philadelphia: G. F. Lasher; 1906.
Note: Illustrated.

Eckels, John H. *Modern Mortuary Science*. Philadelphia: Westbrook Publishing Co.; 1948.

Editors of Consumer Reports. *Funerals: Consumers' Last Rights*. Mount Vernon, NY: Consumers' Union; 1977.

Note: Subtitled *The Consumers' Union Report on Conventional Funerals and Burial...and Some Alternatives, Including Cremation, Direct Cremation, Direct Burial, and Body Donation*. Also published by Pantheon Books, New York.

For annotation, see page 1.

Encyclopedia of Industrial Chemical Analysis, Vol. 12: Embalming Chemicals. New York: John Wiley and Sons; 1971.

Evans, W. E. D. *The Chemistry of Death*. Springfield, IL: Charles C. Thomas; 1963.

Fredrick, Jerome F. *Embalming Problems Caused by Chemotherapeutic Agents*. Boston: Dodge Institute for Advanced Studies; 1968.

Gale, Frederick. *Mortuary Science*. Springfield, IL: Charles C. Thomas; 1960.

A complete textbook on embalming and restorative art. Illustrated with drawings and photographs.

Goldstein, Lynne G. *Mississippian Mortuary Practices*. In *Scientific Paper Series,* No. 4. Kampsville, IL: Center for American Archaeology; 1980.
Note: Illustrated.

Habenstein, Robert W., and William M. Lamers. *Funeral Customs the World Over*. Milwaukee: Bulfin Printing; 1963.
Note: Reprinted in 1974. First printing in 1960.
The authors present a detailed account of mortuary practices throughout the world. Sections are Asia, the Middle East, Africa, Oceania, Europe, Latin America, Canada, and the United States. Especially comprehensive are the chapters on Native American funeral customs and burial rites. Illustrated with photographs. Covers all aspects of funeral service: embalming, burial, funeral coaches and automobiles, funeral homes, mortuary science education, etc. Also includes details on Jewish, Latter-Day Saints, and American Gypsy funeral rites.

Habenstein, Robert W., and William M. Lamers. *The History of American Funeral Directing*. New York: Omnigraphics Inc.; 1990.
Note: Reprint of the original 1955 edition. Second printing in 1956.

For annotation, see page 10.

Havel, James T. *The Kansas State Board of Embalming*. Lawrence, KS: Governmental Research Center, University of Kansas; 1966.

Hinson, Maude R. *Final Report on Literature Search on the Infectious Nature of Dead Bodies for the Embalming Chemical Manufacturers Association*. Cambridge, MA: Embalming

Chemical Manufacturers Association; 1968.
Note: Author was a medical research librarian in Downers Grove, IL.

Jackson, Percival E. *The Law of Cadavers*, 2nd ed. Englewood Cliffs, NJ: Prentice-Hall; 1950.

Johnson, Edward. *A History of the Art and Science of Embalming*. New York: Casket and Sunnyside; 1944.

Kastenbaum, Robert, ed. *Death and Dying*. New York: Arno Press; 1977.
Note: This title constitutes a 40-volume set dealing with all issues surrounding the subjects of death, dying, grief, bereavement, funerals, etc.

Lucas, Alfred. *Ancient Egyptian Materials and Industries*,4th ed. rev. ; 1989.
Note: J. R. Harris, rev.

Martin, Edward. *Psychology of Funeral Service*, 6th ed. Grand Junction, CO: Edward A. Martin.
Note: Third edition published in 1950.
Martin begins the text with a prologue on the importance of education in general and the necessity of mortuary education to society. He discusses a variety of aspects of funeral service from historical background to modern-day practical considerations including an introduction to psychology. He also includes sections on emotion, learning and memory, adjustment to mental conflict, grief, sentiment, religion (with an encyclopedic coverage of 11 religions of the world and 20 religious concepts), funeral rituals (burial, cremation, mutilation, dismemberment, cannibalism, abandonment, and exposure), public relations, and embalming, and a chapter on "psychology in action." Includes the Funeral Service Oath, index, and glossary.

Mayer, J. Sheridan. *Restorative Art*. Philadelphia: Westbrook Publishing Co.; 1943.

Note: Also published by Graphics Arts Press, Livonia, NY, in 1961. Copyright held by Eckels College of Embalming.
Contains detailed instructions on restoring, modeling, and treating all areas of the dead human body. Includes a section on hair restoration. Text also includes a compend of questions at the end of each chapter, bibliography, and index. Illustrated by the author with 566 drawings.

Mayer, Robert G. *Embalming: History, Theory, and Practice*. Norwalk, CT: Appleton and Lange; 1990.

For annotation, see page 2.

McCurdy, C. W. *The Embalmer as Sanitarian: Embalming and Embalming Fluids*. Wooster, OH: University of Wooster; 1895.
Note: Doctoral manuscript.

Mendelsohn, Simon. *Embalming Fluids: Their Historical Development and Formulation, from the Chemical Aspects of the Scientific Art of Preserving Human Remains*. New York: Chemical Publishing Co.; 1940.
Note: Illustrated.

Mitford, Jessica. *The American Way of Death*. New York: Simon and Schuster; 1963.

Note: Also published in 1963 by Fawcett Publications, Greenwich, CT. Numerous reprints.

For annotation, see page 12.

Myers, E. *Textbook on Embalming*. Springfield, OH: Champion Chemical Co.; 1908.

Myers, John. *Manual of Funeral Procedure*. Casper, WY: Prairie Publishing Co.; 1956.

National Funeral Directors Association. *Should the Body Be Present at the Funeral?* Milwaukee: National Funeral Directors Association, 1991.

National Yellow Book of Funeral Directors and Services. Youngstown, OH: Nomis Publications, Inc.

Pervier, Norville Clarence. *A Textbook of Chemistry for Embalmers,* rev. ed. Minneapolis: Burgess Publishing Co.; 1956.
Note: Also published in 1961 by the University of Minnesota, Minneapolis. Illustrated.

Polson, Cyril J., R. P. Brittain, and T. K. Marshall. *Disposal of the Dead.* New York: Philosophical Library; 1953.
Note: Also published in 1962 by English Universities Press, London, and by Charles C. Thomas Publishers, Springfield, IL.
A comprehensive study of burial and cremation practices, focusing somewhat on those of England. Contains a thorough historical introduction to the disposal of the dead. Also includes sections on mediate disposal (death certificates, coroners, registration, etc.), cremation, burial (churchyards, cemeteries, burial grounds), funeral rites, exhumation, embalming, and funeral direction. Distinguishes mummification and embalming as modes of preservation. Treats unusual subjects such as preservation of human heads, ship-burial, and radioactive corpses.

Renouard, C. A., ed. *Taylor's Art of Embalming.* New York: H. E. Taylor and Co.; 1903.

Selzer, Richard. *Mortal Lessons: Notes on the Art of Surgery.* New York: Simon and Schuster; 1974.
Note: Reprinted in 1975 and 1976.
Contains one section of particular interest to funeral service personnel: "The Corpse." The author, with a tongue-in-cheek approach, makes humorous and sarcastic references to embalming, morgues, autopsies, restorative art, and burial. A summary of the text refers to it as "a

rich and stimulating blend of information, reflection, and literary-medical allusion."

Slater, Robert E., ed. *Funeral Service.* Milwaukee: National Funeral Directors Association; 1964.
Note: Author was director of the University of Minnesota's Department of Mortuary Science.

Sleichter, G. M. *A Study of the Combining Activity of Formaldehyde with Tissues.* Cincinnati: University of Cincinnati; 1939.
Note: Thesis submitted in partial fulfillment of the requirements for the master's degree.

Slocum, R. E. *Pre-embalming Considerations.* Boston: Dodge Chemical Co.; 1969.

Sourcebook on Death and Dying, 1st ed. Chicago: Marquis Professional Publications; 1982.

Spriggs, A. O. *The Art and Science of Embalming.* Springfield, OH: Champion Chemical Co.; 1963.
Spriggs, director of service and research for the Champion Company, calls this text the second edition of *The Textbook on Embalming,* the first edition published in 1933. He discusses the history of embalming, arterial embalming, decomposition and preservation, the blood vascular system, body preparation, raising vessels, injection and drainage, cavity treatment, surface tension, discolorations before and after death, diseases of the blood vessels, heart, respiratory system, digestive tract, liver, and kidneys, diabetes, infectious and contagious diseases, heat prostration, preparation of children's bodies, embalming posted bodies, accidental and violent deaths, poisons, health department duties, cause of death, cosmetics, and general anatomy. Includes a compend of 352 questions and answers. Illustrated.

Spriggs, A. O. *Champion Anatomy for Embalmers.* Springfield, OH: Champion Chemical Co.; 1934.
Note: First edition.

A general anatomy text for funeral service professionals with sections on the skeleton, syndesmology, muscles, the nervous system, organs of respiration, organs of digestion, abdominal and pelvic cavity organs, blood, organs of blood circulation, and extensive chapters on blood circulation systems. The author was director of service and research for the Champion Chemical Company at the time of writing. Includes a compend of 350 questions and answers. Indexed.

Spriggs, A. O. *Champion Textbook on Embalming and Anatomy for Embalmers*. Springfield, OH: Champion Chemical Co.; 1944.

Strub, Clarence G. *The Principles and Practices of Embalming*, 4th ed. Dallas: L. G. "Darko" Frederick; 1967.
Note: First edition published under the title *The Theory and Practice of Embalming*; 4th edition also cited as published in 1986.

Walker, J. F. *Formaldehyde*, 3rd ed. In *American Chemical Society Monograph Series*. New York: Reinhold Publishing Corp.; 1964.

Walrath, J., et al. *Formaldehyde Toxicity*. New York: Hemisphere Pub.; 1983.

Epitaphs

Alden, Timothy. *A Collection of American Epitaphs and Inscriptions with Occasional Notes*. New York: Arno Press; 1976.
Note: Two volumes.
A collection of epitaphs with explanatory notes, arranged by state, by city, and alphabetically within cities by persons. A resource on early American attitudes toward death.

Beable, William H. *Epitaphs: Graveyard Humor and Eulogy*. New York: Thomas Y. Crowell; 1925.

Birrell, F. F. L., and F. L. Lucas. *The Art of Dying*. London: L.&V. Woolf; 1930.
Note: Collection of last words.

Cemeteries and Gravemarkers: Voices of American Culture. Ann Arbor, MI: UMI Research Press; 1989.

Culbertson, J., and T. Randall. *Permanent New Yorkers*. Chelsea, VT: Chelsea Green; 1987.

Deacy, William H. *Memorials Today for Tomorrow*. Tate, GA: Georgia Marble Co.; 1928.

Duval, Francis Y., and Ivan B. Rigby. *Early American Gravestone Art in Photographs: Two Hundred Outstanding Examples*. New York: Dover; 1979.

Harris, Paul. *R.I.P.: A Lighthearted Look at Life Through Death*. London: Harrap; 1983.
A humorous, tongue-in-cheek examination of obituaries.

Jordan, Terry G. *Texas Graveyards: A Cultural Legacy*. Austin: University of Texas Press; 1982.

Landau, Elaine. *Death: Everyone's Heritage*. New York: Messner; 1976.
Landau, a librarian, offers a collection of "scraps" and anecdotes about death, suicide, euthanasia, and funerals.

Lindley, Kenneth. *Of Graves and Epitaphs*. London: Hutchison; 1965.

Schafer, Louis S. *Best of Gravestone Humor*. New York: Sterling; 1990.
Note: Illustrated.

Wallis, Charles. *Stories on Stone: A Book of American Epitaphs.* New York: Oxford University Press; 1954.

Ethnic Funeral Customs

Bendann, Effie. *Death Customs: An Analytical Study of Burial Rites.* New York: Alfred A. Knopf; 1930.
Note: Also published in 1930 by Kegan, Paul and Co., London.
An examination of the funerary practices of many nations and religious groups from early times to the date of publication. Discusses the relationship between funerary practices and the belief and thought forms of a people. The text is divided into two sections: similarities and differences (of funeral rites and ceremonies). She covers disposal of the dead, general attitudes toward the corpse, purification, life after death, taboos, mourning, women's connection with funeral rites, totemic conceptions, destruction of property, and cult of the dead. Includes a comprehensive index and glossary of terms.

Buck, Peter, and Te Rangi Hiroa. *Arts and Crafts of Hawaii: Death and Burial.* In *Special Publication Series,* No. 45(13). Honolulu: Bishop Museum; 1957.
Note: Illustrated.

Counts, David and Dorothy, eds. *Coping with the Final Tragedy: Dying and Grieving in Cross Cultural Perspective .* Baywood Publishers; 1991.

Davis, Daniel L. *What to Do When Death Comes.* New York: Federation of Reform Temples.
Discusses Jewish funeral customs and burial rites.

Goldstein, Lynne G. *Mississippian Mortuary Practices.* In *Scientific Paper Series,* No. 4. Kampsville, IL: Center for American Archaeology; 1980.
Note: Illustrated.

Gordon, Anne. *Death Is for the Living.* Edinburgh, Scotland: Paul Harris Publishing; 1984.
Note: Subtitled *The Strange History of Funeral Customs.* While Gordon does make continual references to Scottish funeral rites and customs, the material is nevertheless applicable to American practices. Sections include: coffins, mort bells, funeral hospitality, burial services, mortcloths, walking funerals, hearses, gravestones, mourning, apparel, executions, and body-snatchers. All focus of the superficial aspects of the funeral--as the author claims, death is for the living.

Grollman, Earl A. *Concerning Death: A Practical Guide for the Living.* Boston: Beacon Press; 1974.
Grollman's book is a guide to dealing with the facts and emotions of death. The text contains 20 individually edited sections on the subject of death and funerals--intended primarily for consumers. Pertinent topics covered are: grief, Protestant, Catholic, and Jewish rites, legal concerns, insurance, coroners, funeral directors, cemeteries, memorials (gravemarkers), cremation, organ donation and transplantation, sympathy calls, condolence letters, widows and widowers, suicide, and death education.

Habenstein, Robert W., and William M. Lamers. *Funeral Customs the World Over.* Milwaukee: Bulfin Printing; 1963.
Note: Reprinted in 1974. First printing in 1960.
The authors present a detailed account of mortuary practices throughout the world. Sections are Asia, the Middle East, Africa, Oceania, Europe, Latin America, Canada, and the United States. Especially comprehensive are the chapters on Native American funeral customs and burial rites. Illustrated with photographs. Covers all aspects of funeral service: embalming, burial, funeral coaches and automobiles, funeral homes, mortuary science education, etc. Also includes details on Jewish, Latter-Day Saints, and American Gypsy funeral rites.

Johnson, E. C., and G. R. *Alone in His Glory* [unpublished manuscript on Civil War mortuary practices].

Kastenbaum, Robert, ed. *Death and Dying*. New York: Arno Press; 1977.
Note: This title constitutes a 40-volume set dealing with all issues surrounding the subjects of death, dying, grief, bereavement, funerals, etc.

Kubler-Ross, Elisabeth. *Death: The Final Stage of Growth*. In *Human Development Books: A Series in Applied Behavioral Science*, Joseph and Laurie Braga, general editors, University of Miami Medical School. Englewood Cliffs, NJ: Prentice-Hall; 1975.
Note: Also published in 1974 by Spectrum Books, New York.
A psychiatrist and well-known authority on death, Kubler-Ross discusses many areas surrounding and encompassing the subject. Rites and customs of American Indians, Jews, Hindus, and Buddhists are covered. An essay entitled "Funerals: Time for Grief and Growth" by Roy and Jane Nichols is included.

Lamm, Maurice. *The Jewish Way in Death and Mourning*. New York: Jonathan David Pub.; 1969.

Litten, Julian. *The English Way of Death: The Common Funeral Since 1450*. London: R. Hale; 1991.

Matunde, Skobi. *Crossing the Great River: A Glimpse into the Funeral Rites of African-Americans*. Philadelphia: Freeland Publications; 1990.
Analyzes and discusses a variety of funeral practices of African-Americans. Contains a list of what should be done at the funeral of a loved one. Also gives instructions for preparing a will. Illustrated.

Rabinowicz, Tzvi. *A Guide to Life: Jewish Laws and Customs of Mourning*. Northvale, NJ: Aronson; 1989.

Smith, Donald Kent. *Why Not Cremation*. Philadelphia: Dorrance and Co.; 1970.
Smith begins the short book with the statement, "Today, people who die and are buried can rest assured that eventually their bodies will be dug up, thrown into a common grave, or cremated." With this, he begins his argument for cremation. He includes details on burial and cremation customs and policies in a variety of countries, obtained by writing letters to world embassies.

Tegg, William. *The Last Act: Being the Funeral Rites of Nations and Individuals*. Detroit: Gale Research; 1973.
Note: Reprint.

Types of Funeral Services and Ceremonies. New York: National Association of Colleges of Mortuary Science, Inc.; 1961.

Van Der Zee, James, Owen Dodson, and Camille Billops. *The Harlem Book of the Dead*. New York: Morgan and Morgan; 1978.

Text and poems based on a set of photographs, taken over several years, in a Harlem mortician's funeral parlor. Focuses on African-American funeral rituals, coffins, and remembrance pictures of the deceased. Illustrated.

Exhumation

Davies, M. R. R. *The Law of Burial, Cremation and Exhumation*. London: Shaw and Son; 1956.
Note: Also published by State Mutual Book in 1974.

Gordon, Anne. *Death Is for the Living*. Edinburgh, Scotland: Paul Harris Publishing; 1984.
Note: Subtitled *The Strange History of Funeral Customs*.

While Gordon does make continual references to Scottish funeral rites and customs, the material is nevertheless applicable to American practices. Sections include: coffins, mort bells, funeral hospitality, burial services, mortcloths, walking funerals, hearses, gravestones, mourning, apparel, executions, and body-snatchers. All focus of the superficial aspects of the funeral--as the author claims, death is for the living.

Polson, Cyril J., R. P. Brittain, and T. K. Marshall. *Disposal of the Dead*. New York: Philosophical Library; 1953.
Note: Also published in 1962 by English Universities Press, London, and by Charles C. Thomas Publishers, Springfield, IL.
A comprehensive study of burial and cremation practices, focusing somewhat on those of England. Contains a thorough historical introduction to the disposal of the dead. Also includes sections on mediate disposal (death certificates, coroners, registration, etc.), cremation, burial (churchyards, cemeteries, burial grounds), funeral rites, exhumation, embalming, and funeral direction. Distinguishes mummification and embalming as modes of preservation. Treats unusual subjects such as preservation of human heads, ship-burial, and radioactive corpses.

Floral Decoration

Editors of Consumer Reports. *Funerals: Consumers' Last Rights*. Mount Vernon, NY: Consumers' Union; 1977.

Note: Subtitled *The Consumers' Union Report on Conventional Funerals and Burial...and Some Alternatives, Including Cremation, Direct Cremation, Direct Burial, and Body Donation*. Also published by Pantheon Books, New York.

For annotation, see page 1.

Funeral Tributes II. New York: Art in Flowers Pub. Co.; 1956.
Discusses floral decoration in funeral ceremonies. Illustrated.

Gordon, Anne. *Death Is for the Living.* Edinburgh, Scotland: Paul Harris Publishing; 1984.
Note: Subtitled *The Strange History of Funeral Customs.* While Gordon does make continual references to Scottish funeral rites and customs, the material is nevertheless applicable to American practices. Sections include: coffins, mort bells, funeral hospitality, burial services, mortcloths, walking funerals, hearses, gravestones, mourning, apparel, executions, and body-snatchers. All focus of the superficial aspects of the funeral--as the author claims, death is for the living.

Habenstein, Robert W., and William M. Lamers. *The History of American Funeral Directing.* New York: Omnigraphics Inc.; 1990.
Note: Reprint of the original 1955 edition. Second printing in 1956.

For annotation, see page 10.

Jones, Barbara. *Design for Death.* Indianapolis: Bobbs-Merrill Co.; 1967.
This extensively illustrated book discusses the art, fashion, and design surrounding the subjects of death and funerals. These include: the corpse, shroud, coffin, hearse, "undertaker's shop," floral tributes, the procession, cemetery, crematorium, tomb, and relics and mementos. Filled with historical references and anecdotes.

Funeral Costs

Arvio, Raymond. *The Cost of Dying and What You Can Do About It.* New York: Harper & Row; 1974.

Backman, Allan Earnshaw. *Consumers Look at Burial Practices*. St. Cloud, MN: Council on Consumer Information; 1956.
Discusses concerns and abuses relating to burial practices. Offers suggestions for coping with high prices and high-pressure sales tactics.

Bernard, Hugh Y. *The Law of Death and Disposal of the Dead*. New York: Oceana; 1966.
Discusses the laws and principles surrounding the death of a person including rights and duties of burials, legal problems of funerals, and complications of debts.

Better Business Bureau. *Facts Every Family Should Know About Funerals and Interments*. New York: Better Business Bureau; 1961.

Better Business Bureau. *The Pre-Arrangement and Pre-Financing of Funerals*. New York: Better Business Bureau; 1960.
Note: Also published in 1963 as *Facts You Should Know, Questions You Should Ask About the Pre-Arrangement and Pre-Financing of Funerals*.

Blackwell, Roger D. *Price Levels of Funerals: An Analysis of the Effects of Entry Regulation in a Differentiated Oligopoly*. Evanston, IL: Northwestern University; 1966.
Note: Ph.D. dissertation.

Blackwell, Roger D., W. Wayne Talarzyk, and David C. Beever. *A Manual for the Return-on-Investment Approach to Professional Funeral Pricing*. Columbus, OH: New Horizon; 1976.

Bullough, Vern L. *The Banal and Costly Funeral*. Yellow Springs, OH: The Humanist Association, Humanist House; 1960.
Discusses the laws governing the disposal of the dead and high funeral costs. Also covers mourning as a psychological necessity.

Carlson, Lisa. *Caring for Your Own Dead*. Hinesburg, VT: Upper Access Publishers; 1987.

A complete guide for those who wish to handle funeral arrangements themselves. The text is divided into three parts. Part 1 discusses home funerals, cremation, embalming, burial, body and organ donation, and legal issues. Part 2 lists laws, regulations, and services in each state. Includes names and addresses of state agencies and regulatory boards, crematories, and sites for body donation. The final part is a group of appendixes with details on death certificates, preneed spending, grieving, and consumer information on Federal Trade Commission funeral regulation rules. Also includes glossary of funeral-related terms. Text comes with a sample death certificate from North Dakota.

Cavanaugh, Sally. *How to Bury Your Own Dead in Vermont*. Vermont: Vanguard Press.
This work was the basis and inspiration for Carlson's *Caring for Your Own Dead*.

Douglass, Sam P. *Funeral Homes: Legal and Business Problems*. In *Commercial Law and Practice Course Handbook Series*, edited by Roger A. Needham, "A4-1052"–Number 5. New York: Practising Law Institute; 1971.
This 240-page text, written by the president of Service Corporation International, Houston, TX and chairman of the Practising Law Institute, was prepared for distribution at a workshop with the same title. Subjects and issues covered are: establishing the purchase price, structuring the transaction, advantages and necessity of affiliation in the funeral service industry, representing the seller of a funeral service firm, the product of the funeral director, management planning for growth-oriented firms, public relations, consumer analysis, socioeconomic variables affecting funeral purchase decisions, professional pricing by funeral service firms, and funeral service in the 1970s. Contributors include Howard Raether, Roger D. Blackwell, James F. Engel, Sam J. Lucas, Jr., and B. B. Hollingsworth, Jr. Includes an extended list of those involved in the Practising Law Institute.

Dowd, Quincy L. *Funeral Management and Costs*. Chicago: University of Chicago Press; 1921.
Note: Subtitled *A World Survey of Burial and Cremation*. An early social scientific study of the funeral. Discusses modern development of cremation and witnessing a cremation.

Draznin, Y. *How to Prepare for Death: A Practice Guide*. New York: Hawthorn Books; 1976.
Draznin's guide, published during what she terms "a bibliographic torrent" of major proportion of books on death. She contends in the preface that death has become society's prime nonfictional fascination. Draznin's text is indeed a practical guide, covering all of the details necessary in preparing for your own death: disposing of the body, the mortuary rites, costs, wills, insurance, and estate conservation. She also includes a 10-chapter section on coping with a death in the family, detailing what to do in the cases of sudden death, accidental death, suicide, homicide, and other circumstances. The text also includes an appendix of supplementary reading notes.

Editors of Consumer Reports. *Funerals: Consumers' Last Rights*. Mount Vernon, NY: Consumers' Union; 1977.

Note: Subtitled *The Consumers' Union Report on Conventional Funerals and Burial...and Some Alternatives, Including Cremation, Direct Cremation, Direct Burial, and Body Donation*. Also published by Pantheon Books, New York.

For annotation, see page 1.

Federal Trade Commission Survey of Funeral Prices in the District of Columbia. Washington, DC: Federal Trade Commission, Government Printing Office; 1974.
An extended price study illustrating variations in price and sales practices. A consumer handbook.

Federated Funeral Directors of America. *Analytical Study of Operation Costs and Adult Funeral Sales for 1965*. Chicago: Federated Funeral Directors of America; 1966.

Funeral Directors Institute. *The Great Controversy Relating to Funerals*. Chicago: Funeral Directors Institute.

Gebhart, John C. *Funeral Costs*. New York: G.P. Putnam's Sons; 1928.
Note: Subtitled *What They Average, Are They Too High? Can They Be Reduced?*

Gebhart, John C. *The Reasons for Present-Day Funeral Costs*. New York: G.P. Putnam's Sons; 1927.

Harmer, Ruth Mulvey. *The High Cost of Dying*. New York: Cromwell-Collier Press; 1963.

Hughes, Theodore E., and David Klein. *A Family Guide to Estate Planning, Funeral Arrangements, and Settling an Estate After Death*. New York: Scribner; 1983.

Johannson, Francia Faust, ed. *The Last Rights: A Look at Funerals*. Mills, MD: Owings/Maryland Center for Public Broadcasting; 1975.
Note: Part of the Consumer Survival Kit.
An assortment of articles dealing with funeral planning, cost information, alternatives, and advice.

Jordahl, Edna K. *Planning and Paying for Funerals*. St. Paul, MN: Agricultural Extension Service, University of Minnesota; 1967.
Note: Revised edition. Also cited as written by Edna K. Fordahl in 1969.
This booklet discusses necessary arrangements for funerals, how to plan for them, financing strategies, laws that relate to death, and body donation.

Jowett, Mary W. *A Guide to Funeral Planning*. Independence, MO: Worship Commission, Reorganized Church of Jesus

Christ of Latter Day Saints and Herald Publishing House; 1985.

Krieger, Wilber M. *Successful Funeral Service Management.* Englewood Cliffs, NJ: Prentice-Hall; 1951.
This text is written for both funeral home management personnel and potential funeral professionals. It applies general management principles and concepts to the funeral service business. Krieger discusses how to enter the profession (license requirements, education, personal characteristics, etc.), public attitudes toward funeral service, management responsibilities, selecting a location, setting up the organization, financing, required investments (with furniture, fixtures, and equipment checklists), working capital, attracting business through advertising and other means, merchandising, accounting, forms to use, credits and collections, letter writing, employment policies, personnel relations, and ethics. Includes an appendix of state licensing rules for embalmers and funeral directors compiled by O. J. Willoughby, publisher of *Southern Funeral Director.*

Margolius, Sidney. *Funeral Costs and Death Benefits.* New York: Public Affairs Committee; 1967.

Mitford, Jessica. *The American Way of Death.* New York: Simon and Schuster; 1963.

Note: Also published in 1963 by Fawcett Publications, Greenwich, CT. Numerous reprints.

For annotation, see page 12.

Morgan, Ernest. *A Manual of Death Education and Simple Burial,* 7th ed. Burnsville, NC: Celo Press; 1973.
Note: Published in 1964 as *A Manual for Simple Burial.* Also cited as *Manual of Simple Burial.* The 9th edition was published in 1980. The 10th edition was published in 1984 with the title *Dealing Creatively with Death: A Manual of Death Education and Simple Burial.* The 11th revised

edition, published in 1988, carried this title also.
This booklet discusses efforts at funeral reform during the 1950s and 1960s, suggesting patterns by which, through group interaction, funerals may be made simpler and less costly. Advocates cremation as a means of disposing of the dead, though not exclusively.

Myers, James, Jr. *Cooperative Funeral Associations*. New York: Cooperative League of the U.S.A.; 1946.
Note: Pamphlet 409. Author also cited as James Meyers, Jr. Also cited as published in Chicago.
An early publication indicating some reasons for high-priced funerals and offering suggestions on how consumers might cut costs.

Nora, Fred. *Memorial Associations: What They Are--How They Are Organized*. Chicago: Cooperative League of the USA; 1962.
Discusses how volunteers can organize societies to obtain respectable funerals at decent costs.

Osborne, Ernest. *When You Lose a Loved One*. New York: Public Affairs Committee; 1965.
Note: Pamphlet 269.
A discussion of the emotional, social, and financial problems relating to death in the family.

The Price of Death: A Survey Method and Consumer Guide for Funerals, Cemeteries, and Grave Markers. Washington, DC: US Government Printing Office; 1975.
Note: Consumer Survey Handbook 3. A Federal Trade Commission Publication, Seattle Regional Office.

Raether, Howard C., ed. *The Funeral Director's Practice Management Handbook*. Englewood Cliffs, NJ: Prentice-Hall; 1989.
Essentially a textbook, this title exhaustively examines the subject of funeral home management with contributions from 16 industry leaders, scholars and other experts. Raether divides the text into two parts focusing on personalizing professional funeral service practices and

practice management and marketing for profitable funeral services. Subjects covered include: grief facilitation, education opportunities for funeral personnel, dealing with the clergy, funeral pricing, public relations, personnel, merchandising, pre-need plans, upgrading funeral service facilities, legal concerns, and FTC regulations. Topical issues are treated thoroughly with sections on "Women in Funeral Service," "Computerization of Recordkeeping," and "Regulatory Changes." Illustrated.

Rappaport, Alfred. *An Analysis of Funeral Service Pricing and Quotation Methods.* Milwaukee: National Funeral Directors Association and National Selected Morticians; 1971.

Simpson, Michael A. *The Facts of Death.* Englewood Cliffs, NJ: Spectrum/Prentice-Hall; 1979.
Offers information on how to plan one's funeral and estate and how to avoid unscrupulous funeral directors.

Smith, Curtis A. *Help for the Bereaved: What the Family Should Know.* Chicago: Adams Press; 1972.
Discusses funerals, death certificates, financial affairs, and benefits.

Thomas, Susan. *What to Do, Know and Expect When a Loved One Dies.* Renton, WA: S. K. Thomas; 1984.

United States Senate Committee on the Judiciary. *Antitrust Aspects of the Funeral Industry--Hearings Before the Subcommittee on Antitrust and Monopoly.* Washington, DC: US Government Printing Office; 1964.
A sourcebook on funeral industry economic practices, sales techniques, advertising limits and bans, and pricing. Includes testimony by religious, labor, and consumer leaders as well as by industry officials.

Wilson, Sir Arnold, and Hermann Levy. *Burial Reform and Funeral Costs.* London: Oxford University Press; 1938.

Funeral Directing

American Blue Book of Funeral Directors. New York: Boylston Publications; 1972.

For annotation, see page 20.

Backman, Allan Earnshaw. *Consumers Look at Burial Practices*. St. Cloud, MN: Council on Consumer Information; 1956.
Discusses concerns and abuses relating to burial practices. Offers suggestions for coping with high prices and high-pressure sales tactics.

Basevi, W. H. F. *The Burial of the Dead*. London: George Routledge and Sons; 1920.
A detailed historical, cross-cultural study of burial and cremation from prehistoric times to the twentieth century.

Bishop, John P., and Edmund Wilson. *The Undertaker's Garland*. New York: Haskell House; 1974.

Blackwell, Roger D. , W. Wayne Talarzyk, and David C. Beever. *A Manual for the Return-on-Investment Approach to Professional Funeral Pricing*. Columbus, OH: New Horizon; 1976.

Bowman, Leroy. *The American Funeral: A Study in Guilt, Extravagance, and Sublimity*. Washington, DC: Public Affairs Press; 1959.
Note: Introduction by Harry A. Overstreet. Also published in 1964 by Paperback Library, New York, and reprinted by Greenwood Press in 1973 and 1975.
The first of the contemporary critiques of the funeral written from the viewpoint of a social scientist who sees the funeral as an anachronism in urban society. He advocates using cremation as a means of making a funeral more economical. Discusses the differentiation between the terms "funeral director," "undertaker," and

"mortician." Bowman covers group behavior at funerals, behind-the-scenes activities, family contact with the undertaker, the undertaker's role in the community, and trends in the form and function of funerals.

Cohn, Mike. *Passing the Torch: Transfer Strategies for Your Family Business.* Milwaukee: National Funeral Directors Association; 1990.

Douglass, Sam P. *Funeral Homes: Legal and Business Problems*. In *Commercial Law and Practice Course Handbook Series,* edited by Roger A. Needham, "A4-1052"–Number 5. New York: Practising Law Institute; 1971.
For annotation, see page 68.

Dowd, Quincy L. *Funeral Management and Costs*. Chicago: University of Chicago Press; 1921.
Note: Subtitled *A World Survey of Burial and Cremation.* An early social scientific study of the funeral. Discusses modern development of cremation and witnessing a cremation.

Eckels College of Mortuary Science Inc. *Modern Mortuary Science,* 4th ed. Philadelphia: Westbrook Publishing Co.; 1958.

Feifel, Herman. *The Meaning of Death.* New York: McGraw-Hill; 1959.

Feifel, Herman. *New Meanings of Death.* New York: McGraw-Hill; 1977.

Foran, Eugene F. *Funeral Service Facts and Figures*. Milwaukee: National Funeral Directors Association; 1957.
Note: An annual publication.

Fulton, Robert. *A Compilation of Studies of Attitudes Toward Death, Funerals, Funeral Directors.* [privately printed]; 1967.
Note: Reprinted in 1971 by the Center for Death

Education and Research, University of Minnesota Press, Minneapolis.

Funeral Directors Institute. *The Great Controversy Relating to Funerals*. Chicago: Funeral Directors Institute.

Gale, Frederick. *Mortuary Science*. Springfield, IL: Charles C. Thomas; 1960.

A complete textbook on embalming and restorative art. Illustrated with drawings and photographs.

Gassman, McDill McCown. *Daddy Was an Undertaker*. New York: Vantage Press; 1952.

Gordon, Anne. *Death Is for the Living*. Edinburgh, Scotland: Paul Harris Publishing; 1984.
Note: Subtitled *The Strange History of Funeral Customs*. While Gordon does make continual references to Scottish funeral rites and customs, the material is nevertheless applicable to American practices. Sections include: coffins, mort bells, funeral hospitality, burial services, mortcloths, walking funerals, hearses, gravestones, mourning, apparel, executions, and body-snatchers. All focus of the superficial aspects of the funeral--as the author claims, death is for the living.

Gould, Marilyn. *Communications for Professional Funeral Firm Management*.
Note: Gould is also author of *Adult Manual for Death Education*.

Habenstein, Robert W. *The American Funeral Director* [unpublished doctoral dissertation, University of Chicago, Department of Sociology, 1954] ; 1954.

Habenstein, Robert W., and William M. Lamers. *Funeral Customs the World Over*. Milwaukee: Bulfin Printing; 1963.
Note: Reprinted in 1974. First printing in 1960. The authors present a detailed account of mortuary

practices throughout the world. Sections are Asia, the Middle East, Africa, Oceania, Europe, Latin America, Canada, and the United States. Especially comprehensive are the chapters on Native American funeral customs and burial rites. Illustrated with photographs. Covers all aspects of funeral service: embalming, burial, funeral coaches and automobiles, funeral homes, mortuary science education, etc. Also includes details on Jewish, Latter-Day Saints, and American Gypsy funeral rites.

Habenstein, Robert W., and William M. Lamers. *The History of American Funeral Directing*. New York: Omnigraphics Inc.; 1990.
Note: Reprint of the original 1955 edition. Second printing in 1956.

For annotation, see page 10.

Havel, James T. *The Kansas State Board of Embalming*. Lawrence, KS: Governmental Research Center, University of Kansas; 1966.

Hollingsworth, C. E., and R. O. Pasnau, eds. *The Family in Mourning: A Guide for Health Professionals*. New York: Grune and Stratton; 1977.
This book, although intended for health services personnel, contains information pertinent to funeral directors.

Johnson, E. C., and G. R. *The Undertakers Manual*. Calgary, Alberta: Canadian Funeral News; 1980.

Jones, Barbara. *Design for Death*. Indianapolis: Bobbs-Merrill Co.; 1967.
This extensively illustrated book discusses the art, fashion, and design surrounding the subjects of death and funerals. These include: the corpse, shroud, coffin, hearse, "undertaker's shop," floral tributes, the procession, cemetery, crematorium, tomb, and relics and mementos. Filled with historical references and anecdotes.

Krieger, Wilber M. *Successful Funeral Service Management.* Englewood Cliffs, NJ: Prentice-Hall; 1951.
This text is written for both funeral home management personnel and potential funeral professionals. It applies general management principles and concepts to the funeral service business. Krieger discusses how to enter the profession (license requirements, education, personal characteristics, etc.), public attitudes toward funeral service, management responsibilities, selecting a location, setting up the organization, financing, required investments (with furniture, fixtures, and equipment checklists), working capital, attracting business through advertising and other means, merchandising, accounting, forms to use, credits and collections, letter writing, employment policies, personnel relations, and ethics. Includes an appendix of state licensing rules for embalmers and funeral directors compiled by O. J. Willoughby, publisher of *Southern Funeral Director.*

Kubasak, Mike. *Cremation and the Funeral Director.* Milwaukee: National Funeral Directors Association.

Lamers, William, Jr. *Death, Grief, Mourning, the Funeral and the Child.* Chicago: National Association of Funeral Directors; 1965.

Mitford, Jessica. *The American Way of Death.* New York: Simon and Schuster; 1963.

Note: Also published in 1963 by Fawcett Publications, Greenwich, CT. Numerous reprints.

For annotation, see page 12.

Myers, John. *Manual of Funeral Procedure.* Casper, WY: Prairie Publishing Co.; 1956.

National Association of Funeral Directors. *Manual of Funeral Directing.* London: National Association of Funeral Directors; 1964.

National Funeral Directors Association. *Analysis of Attitudes Toward Funeral Directors*. Milwaukee: National Funeral Directors Association; 1948.

National Funeral Directors Association. *Funeral Home Operations Survey*. Milwaukee: National Funeral Directors Association; 1990.

National Funeral Directors Association. *Resource Manual*. Milwaukee: National Funeral Directors Association; 1979.

National Funeral Directors Association. *Should the Body Be Present at the Funeral?* Milwaukee: National Funeral Directors Association; 1991.

National Yellow Book of Funeral Directors and Services. Youngstown, OH: Nomis Publications, Inc.

Parsons, Talcott. *Findings of the Professional Census*. Milwaukee: National Funeral Directors Association; 1971.

Pine, Vanderlyn R. *Caretaker of the Dead: The American Funeral Director*. New York: Irvington; 1975.
This text is a thorough discussion of the American funeral service professional. It gives a historical portrait of the profession and discusses professionalism, organization within funeral homes, behavior outside the funeral home, public behavior in the funeral home, non-public behavior in the funeral home, the presentation of self, the funeral director's role, caretakers, and personal service. Appendixes include a funeral directing questionnaire, funeral arranger interview schedule, and findings of a professional census. In his introduction, Pine discusses the study of death.

Pine, Vanderlyn R. *Statistical Abstract of Funeral Service Facts and Figures of the United States*. Milwaukee: National Funeral Directors Association; 1990.
Note: Updated versions published regularly.

Porter, W. H., Jr. *The Professional-Commercial Debate: The Funeral Business Trade as a Mirror of Intra-Industry Controversy.* Alliance, OH: Mt. Union College; 1977.

Raether, Howard C. *Funeral Service: A Historical Perspective.* Milwaukee: National Funeral Directors Association; 1990.

Raether, Howard C. *The NFDA Resource Manual.* Milwaukee: National Funeral Directors Association.

Raether, Howard C. *Successful Funeral Service Practice.* Englewood Cliffs, NJ: Prentice-Hall; 1971.

Raether, Howard C., and Robert C. Slater. *The Funeral Director and His Role as a Counselor.* Milwaukee: Bulfin Press; 1975.

Raether, Howard C., ed. *The Funeral Director's Practice Management Handbook.* Englewood Cliffs, NJ: Prentice-Hall; 1989.

Essentially a textbook, this title exhaustively examines the subject of funeral home management with contributions from 16 industry leaders, scholars and other experts. Raether divides the text into two parts focusing on personalizing professional funeral service practices and practice management and marketing for profitable funeral services. Subjects covered include: grief facilitation, education opportunities for funeral personnel, dealing with the clergy, funeral pricing, public relations, personnel, merchandising, pre-need plans, upgrading funeral service facilities, legal concerns, and FTC regulations. Topical issues are treated thoroughly with sections on "Women in Funeral Service," "Computerization of Recordkeeping," and "Regulatory Changes." Illustrated.

Rappaport, Alfred. *An Analysis of Funeral Service Pricing and Quotation Methods.* Milwaukee: National Funeral Directors Association and National Selected Morticians; 1971.

Rudman, Jack. *Funeral Directing Investigator*. In *Career Examination Series*, C-3112. Syosset, NY: National Learning; 1988.

Rudman, Jack. *Mortuary Caretaker*. In *Career Examination Series*, C-500. Syosset, NY: National Learning; 1989.

Rudman, Jack. *Mortuary Technician*. In *Career Examination Series*, C-514. Syosset, NY: National Learning; 1989.

Rudman, Jack. *Senior Mortuary Caretaker*. In *Career Examination Series*, C-721. Syosset, NY: National Learning; 1989.

Slater, Robert E., ed. *Funeral Service*. Milwaukee: National Funeral Directors Association; 1964.
Note: Author was director of the University of Minnesota's Department of Mortuary Science.

Sourcebook on Death and Dying, 1st ed. Chicago: Marquis Professional Publications; 1982.

United States. Department of Commerce. Bureau of the Census. *1982 Census of Service Industries. Preliminary Report. Industry Series. Funeral Services and Crematories*. Washington, DC: Department of Commerce, Bureau of the Census; 1984.

United States Senate Committee on the Judiciary. *Antitrust Aspects of the Funeral Industry--Hearings Before the Subcommittee on Antitrust and Monopoly*. Washington, DC: US Government Printing Office; 1964.
A sourcebook on funeral industry economic practices, sales techniques, advertising limits and bans, and pricing. Includes testimony by religious, labor, and consumer leaders as well as by industry officials.

United States Senate Committee on the Judiciary. *Antitrust Aspects of the Funeral Industry--Views of the Subcommittee on Antitrust and Monopoly*. Washington, DC: US Government Printing Office; 1967.

This document was issued in place of an official report due to differing opinions on the committee. A companion document to the Hearings.

United States Senate. *Preneed Burial Service: Hearings Before the Subcommittee on Frauds and Misrepresentations Affecting the Elderly of Special Committee on Aging, U.S. Senate, 88th Congress, Second Session*. Washington, DC: Government Printing Office; 1964.
Discusses preneed funeral and cemetery purchases for senior citizens. Also reveals and documents fraudulent practices.

Weathers, Neil F. *Dunham's Green Book: Service for the Funeral Directors of New England*, 23rd ed. Wilmot Flat, NH: Dunham Services; 1986.

Wolfelt, Alan D. *Interpersonal Skills Training: A Handbook for Funeral Service Staffs*. Muncie, IN: Accelerated Development Inc.; 1990.
Note: Distributed by the National Funeral Directors Association, Milwaukee. Also cited as *Interpersonal Skills Training: A Handbook for Funeral Home Staffs.*
Wolfelt, a clinical thanatologist and director of the Center for Loss and Life Transition in Fort Collins, CO, presents a how-to text on counseling, grief therapy, and general interpersonal skills training for funeral service personnel. He addresses the need for such training, noting that mortuary science schools focus primarily on anatomy and physiology and little on counseling, a much needed skill for funeral directors, according to Wolfelt. He extensively covers grief, communication techniques, and mourning. Includes an important section on funeral service stress. Also offers list of training opportunities for funeral directors and an index.

Worcester, Alfred. *The Care of the Aged, the Dying, and the Dead*. In *The Literature of Death and Dying Series*. New York: Arno Press; 1950.

Funeral Dress and Apparel

Cunnington, P. *Costume for Births, Marriages, and Deaths.* Atlantic Highlands, NJ: Humanities Press; 1977.

Gorer, Geoffrey. *Death, Grief, and Mourning.* In *The Literature of Death and Dying Series.* New York: Doubleday; 1965.
Note: Published in 1977 by Arno Press, New York.
Though Gorer's text does cover the practicalities of death, funerals, and their aftermath, it is written primarily from from anthropological and psychological viewpoints. Gorer discusses bereavement extensively, with one section devoted to types of bereavement: death of father, death of child, etc. He also covers telling children about death, the afterlife, issues surrounding the clergy and the church, body disposal (the funeral), family gatherings, gravestones, condolences, and mourning. The four appendixes are: current and recent theories of mourning and the present material, a questionnaire with statistical tables, religious beliefs and practices: 1963 and 1950 compared, and the pornography of death. Also includes index of informants quoted more than once.

Habenstein, Robert W., and William M. Lamers. *The History of American Funeral Directing.* New York: Omnigraphics Inc.; 1990.
Note: Reprint of the original 1955 edition. Second printing in 1956.

For annotation, see page 10.

Jones, Barbara. *Design for Death.* Indianapolis: Bobbs-Merrill Co.; 1967.
This extensively illustrated book discusses the art, fashion, and design surrounding the subjects of death and funerals. These include: the corpse, shroud, coffin, hearse, "undertaker's shop," floral tributes, the procession,

cemetery, crematorium, tomb, and relics and mementos. Filled with historical references and anecdotes.

Passing: The Vision of Death in America. Westport, CT: Greenwood Press; 1977.
Contains a history of mortuary customs.

Shelley, Marshall. *Weddings, Funerals, and Special Events, No. 10.* New York: Word Books; 1987.

Van Der Zee, James, Owen Dodson, and Camille Billops. *The Harlem Book of the Dead.* New York: Morgan and Morgan; 1978.

Text and poems based on a set of photographs, taken over several years, in a Harlem mortician's funeral parlor. Focuses on African-American funeral rituals, coffins, and remembrance pictures of the deceased. Illustrated.

Funeral Industry Regulation

American Blue Book of Funeral Directors. New York: Boylston Publications; 1972.

For annotation, see page 20.

Backman, Allan Earnshaw. *Consumers Look at Burial Practices.* St. Cloud, MN: Council on Consumer Information; 1956.
Discusses concerns and abuses relating to burial practices. Offers suggestions for coping with high prices and high-pressure sales tactics.

Blackwell, Roger D. *Price Levels of Funerals: An Analysis of the Effects of Entry Regulation in a Differentiated Oligopoly.* Evanston, IL: Northwestern University; 1966.
Note: Ph.D. dissertation.

Carlson, Lisa. *Caring for Your Own Dead*. Hinesburg, VT: Upper Access Publishers; 1987.
A complete guide for those who wish to handle funeral arrangements themselves. The text is divided into three parts. Part 1 discusses home funerals, cremation, embalming, burial, body and organ donation, and legal issues. Part 2 lists laws, regulations, and services in each state. Includes names and addresses of state agencies and regulatory boards, crematories, and sites for body donation. The final part is a group of appendixes with details on death certificates, preneed spending, grieving, and consumer information on Federal Trade Commission funeral regulation rules. Also includes glossary of funeral-related terms. Text comes with a sample death certificate from North Dakota.

Douglass, Sam P. *Funeral Homes: Legal and Business Problems*. In *Commercial Law and Practice Course Handbook Series*, edited by Roger A. Needham, "A4-1052"--Number 5. New York: Practising Law Institute; 1971.
For annotation, see page 68.

Kahn, Jack E. *United States of America Before Federal Trade Commission: Report of the Presiding Officer on Proposed Trade Regulation Rule Concerning Funeral Industry Practices*. Washington, DC: Federal Trade Commission; 1977.

Mitford, Jessica. *The American Way of Death*. New York: Simon and Schuster; 1963.

Note: Also published in 1963 by Fawcett Publications, Greenwich, CT. Numerous reprints.

For annotation, see page 12.

National Funeral Directors Association. *Federal Trade Commission (FTC) Funeral Rule Compliance Manual*. Milwaukee: National Funeral Directors Association.

National Yellow Book of Funeral Directors and Services. Youngstown, OH: Nomis Publications, Inc.

Neilson, William A. W., and C. Gaylord Watkins. *Proposals for Legislative Reform Aiding the Consumer of Funeral Industry Products and Services.* Burnsville, NC: Celo Press; 1973.
A detailed study of the laws and practices of the United States and Canada relating to funeral arrangements, written for consumers.

Raether, Howard C., ed. *The Funeral Director's Practice Management Handbook.* Englewood Cliffs, NJ: Prentice-Hall; 1989.
Essentially a textbook, this title exhaustively examines the subject of funeral home management with contributions from 16 industry leaders, scholars and other experts. Raether divides the text into two parts focusing on personalizing professional funeral service practices and practice management and marketing for profitable funeral services. Subjects covered include: grief facilitation, education opportunities for funeral personnel, dealing with the clergy, funeral pricing, public relations, personnel, merchandising, pre-need plans, upgrading funeral service facilities, legal concerns, and FTC regulations. Topical issues are treated thoroughly with sections on "Women in Funeral Service," "Computerization of Recordkeeping," and "Regulatory Changes." Illustrated.

United States. Bureau of Consumer Protection. Division of Special Projects. *Funeral Industry Practices: Proposed Trade Regulation Rule and Staff Memorandum.* Washington, DC: Bureau of Consumer Protection, Division of Special Projects; 1975.

United States. Bureau of Consumer Protection. *Funeral Industry Practices: Final Staff Report to the Federal Trade Commission and Proposed Trade Regulation Rule (16 CFR part 453).* Washington, DC: US Government Printing Office; 1978.

United States Senate Committee on the Judiciary. *Antitrust Aspects of the Funeral Industry--Hearings Before the Subcommittee on Antitrust and Monopoly*. Washington, DC: US Government Printing Office; 1964.
A sourcebook on funeral industry economic practices, sales techniques, advertising limits and bans, and pricing. Includes testimony by religious, labor, and consumer leaders as well as by industry officials.

United States Senate Committee on the Judiciary. *Antitrust Aspects of the Funeral Industry--Views of the Subcommittee on Antitrust and Monopoly*. Washington, DC: US Government Printing Office; 1967.
This document was issued in place of an official report due to differing opinions on the committee. A companion document to the Hearings.

United States Senate. *Preneed Burial Service: Hearings Before the Subcommittee on Frauds and Misrepresentations Affecting the Elderly of Special Committee on Aging, U.S. Senate, 88th Congress, Second Session*. Washington, DC: Government Printing Office; 1964.
Discusses preneed funeral and cemetery purchases for senior citizens. Also reveals and documents fraudulent practices.

Watts, Tim J. *The Funeral Industry: Regulating the Disposition of the Dead.* In *Public Administration Series*, P2454. Monticello, IL: Vance Bibliographies; 1988.
The author is public services librarian at Valparaiso University School of Law Library in Valparaiso, IN. The bibliography contains 24 monograph citations and 112 article citations on funeral industry regulation; 13 of the 24 monographs cited are United States government publications. Most of the articles are from law journals or popular periodicals. A general statement on funeral industry regulation precedes the bibliography.

Funeral Laws and Regulations

American Blue Book of Funeral Directors. New York: Boylston Publications; 1972.

For annotation, see page 20.

Backman, Allan Earnshaw. *Consumers Look at Burial Practices*. St. Cloud, MN: Council on Consumer Information; 1956.
Discusses concerns and abuses relating to burial practices. Offers suggestions for coping with high prices and high-pressure sales tactics.

Bernard, Hugh Y. *The Law of Death and Disposal of the Dead*. New York: Oceana; 1966.
Discusses the laws and principles surrounding the death of a person including rights and duties of burials, legal problems of funerals, and complications of debts.

Bullough, Vern L. *The Banal and Costly Funeral*. Yellow Springs, OH: The Humanist Association, Humanist House; 1960.
Discusses the laws governing the disposal of the dead and high funeral costs. Also covers mourning as a psychological necessity.

Carlson, Lisa. *Caring for Your Own Dead*. Hinesburg, VT: Upper Access Publishers; 1987.
A complete guide for those who wish to handle funeral arrangements themselves. The text is divided into three parts. Part 1 discusses home funerals, cremation, embalming, burial, body and organ donation, and legal issues. Part 2 lists laws, regulations, and services in each state. Includes names and addresses of state agencies and regulatory boards, crematories, and sites for body

donation. The final part is a group of appendixes with details on death certificates, preneed spending, grieving, and consumer information on Federal Trade Commission funeral regulation rules. Also includes glossary of funeral-related terms. Text comes with a sample death certificate from North Dakota.

Cavanaugh, Sally. *How to Bury Your Own Dead in Vermont.* Vermont: Vanguard Press.
This work was the basis and inspiration for Carlson's *Caring for Your Own Dead.*

Davies, M. R. R. *The Law of Burial, Cremation and Exhumation.* London: Shaw and Son; 1956.
Note: Also published by State Mutual Book in 1974.

Douglass, Sam P. *Funeral Homes: Legal and Business Problems.* In *Commercial Law and Practice Course Handbook Series,* edited by Roger A. Needham, "A4-1052"--Number 5. New York: Practising Law Institute; 1971.
For annotation, see page 68.

Federal Trade Commission Survey of Funeral Prices in the District of Columbia. Washington, DC: Federal Trade Commission, Government Printing Office; 1974.
An extended price study illustrating variations in price and sales practices. A consumer handbook.

Grollman, Earl A. *Concerning Death: A Practical Guide for the Living.* Boston: Beacon Press; 1974.
Grollman's book is a guide to dealing with the facts and emotions of death. The text contains 20 individually edited sections on the subject of death and funerals--intended primarily for consumers. Pertinent topics covered are: grief, Protestant, Catholic, and Jewish rites, legal concerns, insurance, coroners, funeral directors, cemeteries, memorials (gravemarkers), cremation, organ donation and transplantation, sympathy calls, condolence letters, widows and widowers, suicide, and death education.

Habenstein, Robert W., and William M. Lamers. *The History of American Funeral Directing*. New York: Omnigraphics Inc.; 1990.
Note: Reprint of the original 1955 edition. Second printing in 1956.

For annotation, see page 10.

Jackson, Percival E. *The Law of Cadavers*, 2nd ed. Englewood Cliffs, NJ: Prentice-Hall; 1950.

Jordahl, Edna K. *Planning and Paying for Funerals*. St. Paul, MN: Agricultural Extension Service, University of Minnesota; 1967.
Note: Revised edition. Also cited as written by Edna K. Fordahl in 1969.
This booklet discusses necessary arrangements for funerals, how to plan for them, financing strategies, laws that relate to death, and body donation.

Krieger, Wilber M. *Successful Funeral Service Management*. Englewood Cliffs, NJ: Prentice-Hall; 1951.
This text is written for both funeral home management personnel and potential funeral professionals. It applies general management principles and concepts to the funeral service business. Krieger discusses how to enter the profession (license requirements, education, personal characteristics, etc.), public attitudes toward funeral service, management responsibilities, selecting a location, setting up the organization, financing, required investments (with furniture, fixtures, and equipment checklists), working capital, attracting business through advertising and other means, merchandising, accounting, forms to use, credits and collections, letter writing, employment policies, personnel relations, and ethics. Includes an appendix of state licensing rules for embalmers and funeral directors compiled by O. J. Willoughby, publisher of *Southern Funeral Director*.

Mitford, Jessica. *The American Way of Death.* New York: Simon and Schuster; 1963.

Note: Also published in 1963 by Fawcett Publications, Greenwich, CT. Numerous reprints.

For annotation, see page 12.

National Funeral Directors Association. *Federal Trade Commission (FTC) Funeral Rule Compliance Manual.* Milwaukee: National Funeral Directors Association.

National Funeral Directors Association. *Living with OSHA: Employee Handbook.* Milwaukee: National Funeral Directors Association; 1990.

Neilson, William A. W., and C. Gaylord Watkins. *Proposals for Legislative Reform Aiding the Consumer of Funeral Industry Products and Services.* Burnsville, NC: Celo Press; 1973.
A detailed study of the laws and practices of the United States and Canada relating to funeral arrangements, written for consumers.

New York Times Information Service. *Death and Funeral Practices: [issues and trends].* Parsippany, NJ: NTIS, Inc.; 1978.

Plowe, Mort C., and Rudolph C. Kemppainen. *Funeral Director's Financial Handbook.* Englewood Cliffs, NJ: Prentice-Hall; 1983.
Written by a Michigan funeral director and a Michigan-based business consultant, this handbook offers several tips and suggestions on financial management for funeral professionals. General subjects covered are: managing cash flow records and a procurement system, minimizing payment delays in the probate court system, using an accountant, selecting and utilizing an investment adviser, selecting a business attorney to match professional requirements, protecting assets against personal litigation, professional incorporation, avoiding tax errors,

minimizing risk in planned business expansion, and cost containment for facilities management. Also details funeral home insurance and liability and retirement fund planning. Includes index, sample funeral home purchase record, casket selection diagram, income analysis form, and sample funeral home floor plan.

Raether, Howard C., ed. *The Funeral Director's Practice Management Handbook*. Englewood Cliffs, NJ: Prentice-Hall; 1989.
Essentially a textbook, this title exhaustively examines the subject of funeral home management with contributions from 16 industry leaders, scholars and other experts. Raether divides the text into two parts focusing on personalizing professional funeral service practices and practice management and marketing for profitable funeral services. Subjects covered include: grief facilitation, education opportunities for funeral personnel, dealing with the clergy, funeral pricing, public relations, personnel, merchandising, pre-need plans, upgrading funeral service facilities, legal concerns, and FTC regulations. Topical issues are treated thoroughly with sections on "Women in Funeral Service," "Computerization of Recordkeeping," and "Regulatory Changes." Illustrated.

Sourcebook on Death and Dying, 1st ed. Chicago: Marquis Professional Publications; 1982.

Street, A. L. H. *Street's Mortuary Jurisprudence*.

Stueve, Thomas F. H. *Mortuary Law*. Cincinnati: Cincinnati College of Embalming; 1985.
Note: Publisher also cited as Foundation for Mortuary Education.

United States. Bureau of Consumer Protection. *Funeral Industry Practices: Final Staff Report to the Federal Trade Commission and Proposed Trade Regulation Rule (16 CFR part 453)*. Washington, DC: US Government Printing Office; 1978.

United States Senate Committee on the Judiciary. *Antitrust Aspects of the Funeral Industry--Hearings Before the Subcommittee on Antitrust and Monopoly*. Washington, DC: US Government Printing Office; 1964.
A sourcebook on funeral industry economic practices, sales techniques, advertising limits and bans, and pricing. Includes testimony by religious, labor, and consumer leaders as well as by industry officials.

United States Senate Committee on the Judiciary. *Antitrust Aspects of the Funeral Industry--Views of the Subcommittee on Antitrust and Monopoly*. Washington, DC: US Government Printing Office; 1967.
This document was issued in place of an official report due to differing opinions on the committee. A companion document to the Hearings.

United States Senate. *Preneed Burial Service: Hearings Before the Subcommittee on Frauds and Misrepresentations Affecting the Elderly of Special Committee on Aging, U.S. Senate, 88th Congress, Second Session*. Washington, DC: Government Printing Office; 1964.
Discusses preneed funeral and cemetery purchases for senior citizens. Also reveals and documents fraudulent practices.

Funeral Meditations

Chakour, Charles M. *Brief Funeral Meditations*. Nashville: Abingdon Press; 1971.
Contains a variety of meditations for particular types of funerals: death of a child, suicide, nonbaptized persons, tragic death, etc.

Doyle, Charles H. *Fifty Funeral Homilies*. Christian Classics; 1984.

Poovey, W. A., ed. *Planning a Christian Funeral: A Minister's Guide*. Minneapolis: Augsburg; 1978.
A book of popular funeral sermons with an introductory text on the purpose of a funeral and the facets of the funeral and burial ceremonies. Biblical text accompanies each sermon, and at the end of each the author lists the preacher, the occasion, and comments.

Wallis, Charles Langworthy. *The Funeral Encyclopedia: A Source Book*. Grand Rapids: Baker Book House; 1973.
Note: Originally published in 1953 by Harper and Brothers, New York.
Contains a variety of funeral sermons, poems, prayers, and guidance for the pastor. Text is divided into five sections: the funeral service, a treasury of sermons, an anthology of funeral poems, a sheaf of funeral prayers, and professional conduct. Includes poetry, textual, classification, and topical indexes. All material is attributed to its respective author.

Funeral Music

Archer, H. G. *The Burial Service: Musical Setting*. Philadelphia: General Council Public Board; 1912.

Champlin, Joseph M. *Through Death to Life: Preparing to Celebrate the Funeral Mass*, rev. ed. Notre Dame, IN: Ave Maria; 1990.

Commission on Worship of the Methodist Church. *A Manual for the Funeral*. Nashville: Abingdon Press; 1962.
Contains titles of music especially suited for the organ.

Lovelace, Austin C. *A Collection of Funeral Music*. Nashville: Abingdon Press; 1962.
A collection of musical pieces.

Snell, Frederick A. *Music for Church Funerals and Memorial Services*. Philadelphia: Fortress Press; 1966.
Contains titles of music especially suited for the organ.

Funeral Reform

Arvio, Raymond. *The Cost of Dying and What You Can Do About It.* New York: Harper & Row; 1974.

Backman, Allan Earnshaw. *Consumers Look at Burial Practices*. St. Cloud, MN: Council on Consumer Information; 1956.
Discusses concerns and abuses relating to burial practices. Offers suggestions for coping with high prices and high-pressure sales tactics.

Bowman, Leroy. *The American Funeral: A Study in Guilt, Extravagance, and Sublimity*. Washington, DC: Public Affairs Press; 1959.
Note: Introduction by Harry A. Overstreet. Also published in 1964 by Paperback Library, New York, and reprinted by Greenwood Press in 1973 and 1975.
The first of the contemporary critiques of the funeral written from the viewpoint of a social scientist who sees the funeral as an anachronism in urban society. He advocates using cremation as a means of making a funeral more economical. Discusses the differentiation between the terms "funeral director," "undertaker," and "mortician." Bowman covers group behavior at funerals, behind-the-scenes activities, family contact with the undertaker, the undertaker's role in the community, and trends in the form and function of funerals.

Burgess, Vicki. *The Memorial Societies Movement: A Challenge to the Funeral Industry in the USA*. University Park, PA: Pennsylvania Sociological Society; 1979.

Carlson, Lisa. *Caring for Your Own Dead*. Hinesburg, VT: Upper Access Publishers; 1987.

A complete guide for those who wish to handle funeral arrangements themselves. The text is divided into three parts. Part 1 discusses home funerals, cremation, embalming, burial, body and organ donation, and legal issues. Part 2 lists laws, regulations, and services in each state. Includes names and addresses of state agencies and regulatory boards, crematories, and sites for body donation. The final part is a group of appendixes with details on death certificates, preneed spending, grieving, and consumer information on Federal Trade Commission funeral regulation rules. Also includes glossary of funeral-related terms. Text comes with a sample death certificate from North Dakota.

Cavanaugh, Sally. *How to Bury Your Own Dead in Vermont.* Vermont: Vanguard Press.
This work was the basis and inspiration for Carlson's *Caring for Your Own Dead.*

Continental Association of Funeral and Memorial Societies. *Bibliography of Funeral Reform.* Washington, DC: Continental Association of Funeral and Memorial Societies.
Note: The Continental Association serves as a clearinghouse for information about the nation's memorial societies.

Dempsey, D. *The Way We Die: An Investigation of Death and Dying in America Today.* New York: McGraw-Hill; 1975.

Huntington, R.,and P. Metcalf. *Celebrations of Death: The Anthropology of Mortuary Ritual.* New York: Cambridge University Press; 1979.
Note: Reprinted in 1980, 1981, and 1984. Also cited as *Celebration of Death.* Illustrated.
Particularly important is the conclusion entitled "American Deathways." It includes critiques of the funeral industry.

Mitford, Jessica. *The American Way of Death*. New York: Simon and Schuster; 1963.

Note: Also published in 1963 by Fawcett Publications, Greenwich, CT. Numerous reprints.

For annotation, see page 12.

Morgan, Ernest. *A Manual of Death Education and Simple Burial*, 7th ed. Burnsville, NC: Celo Press; 1973.
Note: Published in 1964 as *A Manual for Simple Burial*. Also cited as *Manual of Simple Burial*. The 9th edition was published in 1980. The 10th edition was published in 1984 with the title *Dealing Creatively with Death: A Manual of Death Education and Simple Burial*. The 11th revised edition, published in 1988, carried this title also.
This booklet discusses efforts at funeral reform during the 1950s and 1960s, suggesting patterns by which, through group interaction, funerals may be made simpler and less costly. Advocates cremation as a means of disposing of the dead, though not exclusively.

National Funeral Directors Association, 21st Century Committee. *Tradition in Transition*. Milwaukee: National Funeral Directors Association; 1981.

New York Times Information Service. *Death and Funeral Practices: [issues and trends]*. Parsippany, NJ: NTIS, Inc.; 1978.

Simpson, Michael A. *The Facts of Death*. Englewood Cliffs, NJ: Spectrum/Prentice-Hall; 1979.
Offers information on how to plan one's funeral and estate and how to avoid unscrupulous funeral directors.

United States. Department of Housing and Urban Development. *Commemorative Parks from Abandoned Public Cemeteries*. Washington, DC: US Government Printing Office; 1971.

United States Senate Committee on the Judiciary. *Antitrust Aspects of the Funeral Industry--Hearings Before the Subcommittee on Antitrust and Monopoly*. Washington, DC: US Government Printing Office; 1964.
A sourcebook on funeral industry economic practices, sales techniques, advertising limits and bans, and pricing. Includes testimony by religious, labor, and consumer leaders as well as by industry officials.

United States Senate Committee on the Judiciary. *Antitrust Aspects of the Funeral Industry--Views of the Subcommittee on Antitrust and Monopoly*. Washington, DC: US Government Printing Office; 1967.
This document was issued in place of an official report due to differing opinions on the committee. A companion document to the Hearings.

Wilson, Sir Arnold, and Hermann Levy. *Burial Reform and Funeral Costs*. London: Oxford University Press; 1938.

Funeral Service Management

Blackwell, Roger D. , W. Wayne Talarzyk, and David C. Beever. *A Manual for the Return-on-Investment Approach to Professional Funeral Pricing*. Columbus, OH: New Horizon; 1976.

Cohn, Mike. *Passing the Torch: Transfer Strategies for Your Family Business*. Milwaukee: National Funeral Directors Association.

Douglass, Sam P. *Funeral Homes: Legal and Business Problems*. In *Commercial Law and Practice Course Handbook Series*, edited by Roger A. Needham, "A4-1052"--Number 5. New York: Practising Law Institute; 1971.
For annotation, see page 68.

Dowd, Quincy L. *Funeral Management and Costs*. Chicago: University of Chicago Press; 1921.
Note: Subtitled *A World Survey of Burial and Cremation.* An early social scientific study of the funeral. Discusses modern development of cremation and witnessing a cremation.

Federal Civil Defense Administration. *Mortuary Science in Civil Defense* (TM-11-12 [Technical Manual]). Washington, DC: Government Printing Office; 1956.
Discusses the establishment of civil defense mortuary services. Disaster plans, staffing and management concerns, training staff, identification of the dead, and interment are covered. Includes mortuary services flow chart illustrating communication and body transport patterns in a disaster setting.

Federated Funeral Directors of America. *Analytical Study of Operation Costs and Adult Funeral Sales for 1965*. Chicago: Federated Funeral Directors of America; 1966.

Gale, Frederick. *Mortuary Science*. Springfield, IL: Charles C. Thomas; 1960.

A complete textbook on embalming and restorative art. Illustrated with drawings and photographs.

Gould, Marilyn. *Communications for Professional Funeral Firm Management*.
Note: Gould is also author of *Adult Manual for Death Education*.

Habenstein, Robert W., and William M. Lamers. *The History of American Funeral Directing*. New York: Omnigraphics Inc.; 1990.
Note: Reprint of the original 1955 edition. Second printing in 1956.

For annotation, see page 10.

Herman, Roger E. *Keeping Good People: Strategies for Solving the Dilemma of the Decade*. Milwaukee: National Funeral Directors Association.

Krieger, Wilber M. *Successful Funeral Service Management*. Englewood Cliffs, NJ: Prentice-Hall; 1951.
This text is written for both funeral home management personnel and potential funeral professionals. It applies general management principles and concepts to the funeral service business. Krieger discusses how to enter the profession (license requirements, education, personal characteristics, etc.), public attitudes toward funeral service, management responsibilities, selecting a location, setting up the organization, financing, required investments (with furniture, fixtures, and equipment checklists), working capital, attracting business through advertising and other means, merchandising, accounting, forms to use, credits and collections, letter writing, employment policies, personnel relations, and ethics. Includes an appendix of state licensing rules for embalmers and funeral directors compiled by O. J. Willoughby, publisher of *Southern Funeral Director*.

Market for Funeral and Cremation Services. New York: Business Trends; 1985.

Marks, Amy Seidel, and Bobby J. Calder. *Attitudes Toward Death and Funerals* . Evanston, IL: The Center for Marketing Sciences, J. L. Kellogg Graduate School of Management, Northwestern University; 1982.
Note: Place of publication also cited incorrectly in some publications as Evansville, IL.

National Funeral Directors Association. *Living with OSHA: Employee Handbook*. Milwaukee: National Funeral Directors Association; 1990.

Nichols, Charles H., Howard C. Raether, Thomas H. Clark, and Vanderlyn R. Pine, eds. *How to Evaluate Your Funeral*

Home. Evanston, IL: National Foundation of Funeral Service; 1976.

Pine, Vanderlyn R. *Caretaker of the Dead: The American Funeral Director*. New York: Irvington; 1975.
This text is a thorough discussion of the American funeral service professional. It gives a historical portrait of the profession and discusses professionalism, organization within funeral homes, behavior outside the funeral home, public behavior in the funeral home, non-public behavior in the funeral home, the presentation of self, the funeral director's role, caretakers, and personal service. Appendixes include a funeral directing questionnaire, funeral arranger interview schedule, and findings of a professional census. In his introduction, Pine discusses the study of death.

Pine, Vanderlyn R. *Statistical Abstract of Funeral Service Facts and Figures of the United States*. Milwaukee: National Funeral Directors Association; 1990.
Note: Updated versions published regularly.

Plowe, Mort C., and Rudolph C. Kemppainen. *Funeral Director's Financial Handbook*. Englewood Cliffs, NJ: Prentice-Hall; 1983.
Written by a Michigan funeral director and a Michigan-based business consultant, this handbook offers several tips and suggestions on financial management for funeral professionals. General subjects covered are: managing cash flow records and a procurement system, minimizing payment delays in the probate court system, using an accountant, selecting and utilizing an investment adviser, selecting a business attorney to match professional requirements, protecting assets against personal litigation, professional incorporation, avoiding tax errors, minimizing risk in planned business expansion, and cost containment for facilities management. Also details funeral home insurance and liability and retirement fund planning. Includes index, sample funeral home purchase

record, casket selection diagram, income analysis form, and sample funeral home floor plan.

Porter, W. H., Jr. *The Professional-Commercial Debate: The Funeral Business Trade as a Mirror of Intra-Industry Controversy.* Alliance, OH: Mt. Union College; 1977.

Raether, Howard C. *The NFDA Resource Manual.* Milwaukee: National Funeral Directors Association.

Raether, Howard C., ed. *The Funeral Director's Practice Management Handbook.* Englewood Cliffs, NJ: Prentice-Hall; 1989.
Essentially a textbook, this title exhaustively examines the subject of funeral home management with contributions from 16 industry leaders, scholars and other experts. Raether divides the text into two parts focusing on personalizing professional funeral service practices and practice management and marketing for profitable funeral services. Subjects covered include: grief facilitation, education opportunities for funeral personnel, dealing with the clergy, funeral pricing, public relations, personnel, merchandising, pre-need plans, upgrading funeral service facilities, legal concerns, and FTC regulations. Topical issues are treated thoroughly with sections on "Women in Funeral Service," "Computerization of Recordkeeping," and "Regulatory Changes." Illustrated.

Rappaport, Alfred. *An Analysis of Funeral Service Pricing and Quotation Methods.* Milwaukee: National Funeral Directors Association and National Selected Morticians; 1971.

United States Senate Committee on the Judiciary. *Antitrust Aspects of the Funeral Industry--Hearings Before the Subcommittee on Antitrust and Monopoly.* Washington, DC: US Government Printing Office; 1964.
A sourcebook on funeral industry economic practices, sales techniques, advertising limits and bans, and pricing.

Includes testimony by religious, labor, and consumer leaders as well as by industry officials.

United States Senate Committee on the Judiciary. *Antitrust Aspects of the Funeral Industry--Views of the Subcommittee on Antitrust and Monopoly*. Washington, DC: US Government Printing Office; 1967.
This document was issued in place of an official report due to differing opinions on the committee. A companion document to the Hearings.

Ward, A. E., ed. *Study Courses in Funeral Management*. London: National Association of Funeral Directors; 1948.

Weathers, Neil F. *Dunham's Green Book: Service for the Funeral Directors of New England*, 23rd ed. Wilmot Flat, NH: Dunham Services; 1986.

Wolfelt, Alan D. *Interpersonal Skills Training: A Handbook for Funeral Service Staffs*. Muncie, IN: Accelerated Development Inc.; 1990.
Note: Distributed by the National Funeral Directors Association, Milwaukee. Also cited as *Interpersonal Skills Training: A Handbook for Funeral Home Staffs.*
Wolfelt, a clinical thanatologist and director of the Center for Loss and Life Transition in Fort Collins, CO, presents a how-to text on counseling, grief therapy, and general interpersonal skills training for funeral service personnel. He addresses the need for such training, noting that mortuary science schools focus primarily on anatomy and physiology and little on counseling, a much needed skill for funeral directors, according to Wolfelt. He extensively covers grief, communication techniques, and mourning. Includes an important section on funeral service stress. Also offers list of training opportunities for funeral directors and an index.

Funeral Services

Allen, R. Earl. *Funeral Source Book.* In *Preaching Helps Series.* Grand Rapids: Baker Books; 1984.

Anders, Rebecca. *A Look at Death.* Minneapolis: Lerner; 1977.
Note: Reprinted in 1984.
A book intended to help understand death and funeral customs.

Armstrong, H. G. *The American Way of Dying.* Hicksville, NY: Exposition Press; 1978.

Bendann, Effie. *Death Customs: An Analytical Study of Burial Rites.* New York: Alfred A. Knopf; 1930.
Note: Also published in 1930 by Kegan, Paul and Co., London.
An examination of the funerary practices of many nations and religious groups from early times to the date of publication. Discusses the relationship between funerary practices and the belief and thought forms of a people. The text is divided into two sections: similarities and differences (of funeral rites and ceremonies). She covers disposal of the dead, general attitudes toward the corpse, purification, life after death, taboos, mourning, women's connection with funeral rites, totemic conceptions, destruction of property, and cult of the dead. Includes a comprehensive index and glossary of terms.

Berrill, Margaret. *Mummies, Masks, and Mourners.* New York: Dutton; 1990.

Better Business Bureau. *Facts Every Family Should Know About Funerals and Interments.* New York: Better Business Bureau; 1961.

Bishop, John P., and Edmund Wilson. *The Undertaker's Garland.* New York: Haskell House; 1974.

Bowman, Leroy. *The American Funeral: A Study in Guilt, Extravagance, and Sublimity.* Washington, DC: Public Affairs Press; 1959.

Note: Introduction by Harry A. Overstreet. Also published in 1964 by Paperback Library, New York, and reprinted by Greenwood Press in 1973 and 1975.
The first of the contemporary critiques of the funeral written from the viewpoint of a social scientist who sees the funeral as an anachronism in urban society. He advocates using cremation as a means of making a funeral more economical. Discusses the differentiation between the terms "funeral director," "undertaker," and "mortician." Bowman covers group behavior at funerals, behind-the-scenes activities, family contact with the undertaker, the undertaker's role in the community, and trends in the form and function of funerals.

Champlin, Joseph M. *Through Death to Life: Preparing to Celebrate the Funeral Mass*, rev. ed. Notre Dame, IN: Ave Maria; 1990.

Christensen, James L. *Complete Funeral Manual*. New York: Revell; 1967.

Christensen, James L. *Funeral Services*. New York: Revell; 1959.

Christensen, James L. *Funeral Services for Today*. New York: Revell; 1977.

Coleman, Reverend William L. *It's Your Funeral*. Wheaton, IL: Tyndale House; 1979.
Coleman, a Christian pastor, encourages advance planning of death arrangements and provides a general overview of some available options.

Davis, Daniel L. *What to Do When Death Comes*. New York: Federation of Reform Temples.
Discusses Jewish funeral customs and burial rites.

Dempsey, D. *The Way We Die: An Investigation of Death and Dying in America Today*. New York: McGraw-Hill; 1975.

Dickerson, Robert B. *Final Placement: A Guide to Deaths, Funerals, and Burials of Notable Americans.* Algonac, MI: Reference Publications; 1982.
Note: Keith Irvine, series editor.

Eaton, Hubert. *The Comemoral.* Los Angeles: Academy Press; 1954.

Eckels, John H. *Modern Mortuary Science.* Philadelphia: Westbrook Publishing Co.; 1948.

Editors of Consumer Reports. *Funerals: Consumers' Last Rights.* Mount Vernon, NY: Consumers' Union; 1977.

Note: Subtitled *The Consumers' Union Report on Conventional Funerals and Burial...and Some Alternatives, Including Cremation, Direct Cremation, Direct Burial, and Body Donation.* Also published by Pantheon Books, New York.

For annotation, see page 1.

Engram, Sara. *Mortal Matters: When a Loved One Dies.* Kansas City, MO: Andrews and McMeel; 1990.

Facts Every Family Should Know, 3rd ed. Forest Park, IL: Wilbert, Inc.; 1967.
Note: First and second editions published in 1960 and 1964, respectively, by Wilbert W. Haase Co., Forest Park, IL. Illustrated with burial vault information.
This short booklet, published by a burial vault manufacturer, contains information for consumers on funeral customs, burial vaults, making a will, and social security benefits, and veterans' benefits, including burial in national cemeteries. Also contains a section for recording family information and final instructions regarding funeral arrangements.

Farrell, James J. *Inventing the American Way of Death, 1830-1920.* In *American Civilization Series,* Allen F. Davis.

Philadelphia: Temple University Press; 1980.
Farrell contends that death is a cultural event and that societies reveal themselves in their treatment of death. Significant sections include the development of the modern cemetery, the modernization of funeral service, and the cosmological contexts of death. Farrell describes and analyzes the development of the American way of death. It is not like Jessica Mitford's *The American Way of Death,*which focuses on the funeral industry's profit motive. This text emphasizes the complexity of cultural change.

Fulton, Robert, and R. Bendiksen. *Death and Identity*. Bowie, MD: Charles Press; 1976.
Note: Revised edition.

Gale, Frederick. *Mortuary Science*. Springfield, IL: Charles C. Thomas; 1960.

A complete textbook on embal ming and restorative art. Illustrated with drawings and photographs.

Garrison, Webb B. *Strange Facts About Death.* Nashville: Abingdon Press; 1978.

Habenstein, Robert W., and William M. Lamers. *Funeral Customs the World Over*. Milwaukee: Bulfin Printing; 1963.
Note: Reprinted in 1974. First printing in 1960.
The authors present a detailed account of mortuary practices throughout the world. Sections are Asia, the Middle East, Africa, Oceania, Europe, Latin America, Canada, and the United States. Especially comprehensive are the chapters on Native American funeral customs and burial rites. Illustrated with photographs. Covers all aspects of funeral service: embalming, burial, funeral coaches and automobiles, funeral homes, mortuary science education, etc. Also includes details on Jewish, Latter-Day Saints, and American Gypsy funeral rites.

Habenstein, Robert W., and William M. Lamers. *The History of American Funeral Directing*. New York: Omnigraphics Inc.; 1990.
Note: Reprint of the original 1955 edition. Second printing in 1956.

For annotation, see page 10.

Harmer, Ruth Mulvey. *The High Cost of Dying*. New York: Cromwell-Collier Press; 1963.

Hendin, David. *Death as a Fact of Life*. New York: W. W. Norton; 1973.
The author, a former medical science journalist, offers a compendium of serious information on death, from both a scientific and pop culture perspective. He recommends cremation as a way of dealing with the land-use crisis. Hendin also suggests transforming cemeteries into playgrounds.

Irion, Paul E. *The Funeral: An Experience of Value*. Lancaster, PA: Theological Seminary; 1956.
Note: Also cited as a publication of the National Funeral Directors Association, Milwaukee.

Irion, Paul E. *The Funeral: Vestige or Value?* In *The Literature of Death and Dying Series*. New York: Arno Press; 1966.
Note: Also published with Abingdon Press, Nashville, in the same year. Reprinted by Arno Press in 1976.
Based on religious, cultural, social, and psychological understanding of the nature of the funeral. Contemporary practices are evaluated in light of the valuable functions of the funeral, and new designs are proposed to conserve significant values. Contains sections on cremation and memorial societies.

Irion, Paul E. *The Funeral and the Mourners*. Nashville: Abingdon Press; 1954.
Note: Author also cited as "Ernest F. Irion" for this title.

Irion, Paul E. *A Manual and Guide for Those Who Conduct a Humanist Funeral Service.* Baltimore: Waverly Press; 1971.

Jackson, Edgar N. *The Christian Funeral.* New York: Channel Press; 1966.
An analysis of the religious significance of the funeral from a Christian perspective with special emphasis on the funeral mediation.

Jackson, Edgar N. *The Significance of the Christian Funeral.* Milwaukee: National Funeral Directors Association; 1966.

Johannson, Francia Faust, ed. *The Last Rights: A Look at Funerals.* Mills, MD: Owings/Maryland Center for Public Broadcasting; 1975.
Note: Part of the Consumer Survival Kit.
An assortment of articles dealing with funeral planning, cost information, alternatives, and advice.

Johnson, J. and M. *Tell Me Papa: A Family Book for Children's Questions About Death and Funerals.* Council Bluffs, IA: Centering Corporation; 1978.
Details for children on hearses, caskets, graves, and vaults. Illustrated by Shari Borum.

Jones, E. Ray. *Funeral Manual.* Cincinnati: Standard Pub.; 1991.

Jowett, Mary W. *A Guide to Funeral Planning.* Independence, MO: Worship Commission, Reorganized Church of Jesus Christ of Latter Day Saints and Herald Publishing House; 1985.

Kastenbaum, Robert, ed. *Death and Dying.* New York: Arno Press; 1977.
Note: This title constitutes a 40-volume set dealing with all issues surrounding the subjects of death, dying, grief, bereavement, funerals, etc.

Lamers, William, Jr. *Death, Grief, Mourning, the Funeral and the Child*. Chicago: National Association of Funeral Directors; 1965.

Lamm, Maurice. *The Jewish Way in Death and Mourning*. New York: Jonathan David Pub.; 1969.

Lamont, Corliss. *A Humanist Funeral Service*. New York: Horizon Press; 1954.
Note: Originally published in 1947 by Beacon Press. New revised edition (also 3rd edition) published in 1977 by Prometheus Books, Buffalo, NY.

Landau, Elaine. *Death: Everyone's Heritage*. New York: Messner; 1976.
Landau, a librarian, offers a collection of "scraps" and anecdotes about death, suicide, euthanasia, and funerals.

Litten, Julian. *The English Way of Death: The Common Funeral Since 1450*. London: R. Hale; 1991.

Margolis, Otto S. *Grief and the Meaning of the Funeral*. Edison, NJ: Mss Information Corporation; 1975.
Note: Also cited as published in New York.

Matunde, Skobi. *Crossing the Great River: A Glimpse into the Funeral Rites of African-Americans*. Philadelphia: Freeland Publications; 1990.
Analyzes and discusses a variety of funeral practices of African-Americans. Contains a list of what should be done at the funeral of a loved one. Also gives instructions for preparing a will. Illustrated.

Mitford, Jessica. *The American Way of Death*. New York: Simon and Schuster; 1963.

Note: Also published in 1963 by Fawcett Publications, Greenwich, CT. Numerous reprints.

For annotation, see page 12.

Morgan, Ernest. *A Manual of Death Education and Simple Burial*, 7th ed. Burnsville, NC: Celo Press; 1973.
Note: Published in 1964 as *A Manual for Simple Burial*. Also cited as *Manual of Simple Burial*. The 9th edition was published in 1980. The 10th edition was published in 1984 with the title *Dealing Creatively with Death: A Manual of Death Education and Simple Burial*. The 11th revised edition, published in 1988, carried this title also.
This booklet discusses efforts at funeral reform during the 1950s and 1960s, suggesting patterns by which, through group interaction, funerals may be made simpler and less costly. Advocates cremation as a means of disposing of the dead, though not exclusively.

Mosman, B. C., and M. W. Stark. *The Last Salute: Civil and Military Funerals, 1921-1969*. Washington, DC: Department of the Army; 1971.

National Funeral Directors Association. *Should the Body Be Present at the Funeral?* Milwaukee: National Funeral Directors Association.

New York Times Information Service. *Death and Funeral Practices: [issues and trends]*. Parsippany, NJ: NTIS, Inc.; 1978.

Passing: The Vision of Death in America. Westport, CT: Greenwood Press; 1977.
Contains a history of mortuary customs.

Pine, Vanderlyn R., et al. *Acute Grief and the Funeral*. Springfield, IL: Charles C. Thomas; 1976.
Note: Other contributing editors are A. Kutscher, D. Peretz, R. Slater, R. DeBellis, R. Volk, and D. Cherico.

Raether, Howard C. *Funeral Service: A Historical Perspective*. Milwaukee: National Funeral Directors Association; 1990.

Raether, Howard C., and Robert C. Slater. *The Funeral: Facing Death as an Experience of Life*. Milwaukee: National Funeral Directors Association; 1974.

Rush, Alfred C. *Death and Burial in Christian Antiquity*. Washington, DC: Catholic University of America Press; 1941.

Shelley, Marshall. *Weddings, Funerals, and Special Events, No. 10*. New York: Word Books; 1987.

Simpson, Michael A. *The Facts of Death*. Englewood Cliffs, NJ: Spectrum/Prentice-Hall; 1979.
Offers information on how to plan one's funeral and estate and how to avoid unscrupulous funeral directors.

Slater, Robert E., ed. *Funeral Service*. Milwaukee: National Funeral Directors Association; 1964.
Note: Author was director of the University of Minnesota's Department of Mortuary Science.

Smith, Curtis A. *Help for the Bereaved: What the Family Should Know*. Chicago: Adams Press; 1972.
Discusses funerals, death certificates, financial affairs, and benefits.

Sourcebook on Death and Dying, 1st ed. Chicago: Marquis Professional Publications; 1982.

Tegg, William. *The Last Act: Being the Funeral Rites of Nations and Individuals*. Detroit: Gale Research; 1973.
Note: Reprint.

Thomas, Susan. *What to Do, Know and Expect When a Loved One Dies*. Renton, WA: S. K. Thomas; 1984.

Types of Funeral Services and Ceremonies. New York: National Association of Colleges of Mortuary Science, Inc.; 1961.

United States. Department of Commerce. Bureau of the Census. *1982 Census of Service Industries. Preliminary Report. Industry Series. Funeral Services and Crematories.* Washington, DC: Department of Commerce, Bureau of the Census; 1984.

University of Wisconsin. *Wisconsin Funeral Service: A Consumer's Guide*, 3rd ed. Madison, WI: University of Wisconsin; 1987.

Wallis, Charles Langworthy. *The Funeral Encyclopedia: A Source Book.* Grand Rapids: Baker Book House; 1973.
Note: Originally published in 1953 by Harper and Brothers, New York.
Contains a variety of funeral sermons, poems, prayers, and guidance for the pastor. Text is divided into five sections: the funeral service, a treasury of sermons, an anthology of funeral poems, a sheaf of funeral prayers, and professional conduct. Includes poetry, textual, classification, and topical indexes. All material is attributed to its respective author.

Wesner, Maralene and Miles. *A Time to Weep: Funeral and Grief Messages.* Idabel, OK: Diversity OKLA; 1988.

Worcester, Alfred. *The Care of the Aged, the Dying, and the Dead.* In *The Literature of Death and Dying Series*. New York: Arno Press; 1950.

Funeral Transportation

Bayley, Joseph. *The View from the Hearse*. Elgin, IL: David D. Cooke; 1969.

Gordon, Anne. *Death Is for the Living*. Edinburgh, Scotland: Paul Harris Publishing; 1984.
Note: Subtitled *The Strange History of Funeral Customs.* While Gordon does make continual references to Scottish funeral rites and customs, the material is nevertheless applicable to American practices. Sections include:

coffins, mort bells, funeral hospitality, burial services, mortcloths, walking funerals, hearses, gravestones, mourning, apparel, executions, and body-snatchers. All focus of the superficial aspects of the funeral--as the author claims, death is for the living.

Habenstein, Robert W., and William M. Lamers. *Funeral Customs the World Over*. Milwaukee: Bulfin Printing; 1963.
Note: Reprinted in 1974. First printing in 1960.
The authors present a detailed account of mortuary practices throughout the world. Sections are Asia, the Middle East, Africa, Oceania, Europe, Latin America, Canada, and the United States. Especially comprehensive are the chapters on Native American funeral customs and burial rites. Illustrated with photographs. Covers all aspects of funeral service: embalming, burial, funeral coaches and automobiles, funeral homes, mortuary science education, etc. Also includes details on Jewish, Latter-Day Saints, and American Gypsy funeral rites.

Habenstein, Robert W., and William M. Lamers. *The History of American Funeral Directing*. New York: Omnigraphics Inc.; 1990.
Note: Reprint of the original 1955 edition. Second printing in 1956.

For annotation, see page 10.

Johnson, J. and M. *Tell Me Papa: A Family Book for Children's Questions About Death and Funerals*. Council Bluffs, IA: Centering Corporation; 1978.
Details for children on hearses, caskets, graves, and vaults. Illustrated by Shari Borum.

Jones, Barbara. *Design for Death*. Indianapolis: Bobbs-Merrill Co.; 1967.
This extensively illustrated book discusses the art, fashion, and design surrounding the subjects of death and funerals. These include: the corpse, shroud, coffin, hearse, "undertaker's shop," floral tributes, the procession,

cemetery, crematorium, tomb, and relics and mementos. Filled with historical references and anecdotes.

Martin, Edward. *Hearses and Funeral Cars* [unpublished manuscript]; 1947.
Note: Title held in the archives of the National Funeral Directors Association in Milwaukee, WI.

McPherson, Thomas A. *American Funeral Cars and Ambulances Since 1900.* In the *Automotive Series,* George H. Dammann, editor. Glen Ellyn, IL: Crestline Pub.; 1973.
Note: Illustrated.

Gravemarkers

Alden, Timothy. *A Collection of American Epitaphs and Inscriptions with Occasional Notes.* New York: Arno Press; 1976.
Note: Two volumes.
A collection of epitaphs with explanatory notes, arranged by state, by city, and alphabetically within cities by persons. A resource on early American attitudes toward death.

Beable, William H. *Epitaphs: Graveyard Humor and Eulogy.* New York: Thomas Y. Crowell; 1925.

Cemeteries and Gravemarkers: Voices of American Culture. Ann Arbor, MI: UMI Research Press; 1989.

Culbertson, J., and T. Randall. *Permanent New Yorkers.* Chelsea, VT: Chelsea Green; 1987.

Deacy, William H. *Memorials Today for Tomorrow.* Tate, GA: Georgia Marble Co.; 1928.

Duval, Francis Y., and Ivan B. Rigby. *Early American Gravestone Art in Photographs: Two Hundred Outstanding Examples.* New York: Dover; 1979.

Forest Lawn Memorial-Park Association. *Art Guide of Forest Lawn*. Los Angeles: Forest Lawn Memorial-Park Association; 1956.

Gordon, Anne. *Death Is for the Living*. Edinburgh, Scotland: Paul Harris Publishing; 1984.
Note: Subtitled *The Strange History of Funeral Customs*. While Gordon does make continual references to Scottish funeral rites and customs, the material is nevertheless applicable to American practices. Sections include: coffins, mort bells, funeral hospitality, burial services, mortcloths, walking funerals, hearses, gravestones, mourning, apparel, executions, and body-snatchers. All focus of the superficial aspects of the funeral--as the author claims, death is for the living.

Gorer, Geoffrey. *Death, Grief, and Mourning*. In *The Literature of Death and Dying Series*. New York: Doubleday; 1965.
Note: Published in 1977 by Arno Press, New York. Though Gorer's text does cover the practicalities of death, funerals, and their aftermath, it is written primarily from from anthropological and psychological viewpoints. Gorer discusses bereavement extensively, with one section devoted to types of bereavement: death of father, death of child, etc. He also covers telling children about death, the afterlife, issues surrounding the clergy and the church, body disposal (the funeral), family gatherings, gravestones, condolences, and mourning. The four appendixes are: current and recent theories of mourning and the present material, a questionnaire with statistical tables, religious beliefs and practices: 1963 and 1950 compared, and the pornography of death. Also includes index of informants quoted more than once.

Grollman, Earl A. *Concerning Death: A Practical Guide for the Living*. Boston: Beacon Press; 1974.
Grollman's book is a guide to dealing with the facts and emotions of death. The text contains 20 individually edited sections on the subject of death and funerals--intended primarily for consumers. Pertinent topics

covered are: grief, Protestant, Catholic, and Jewish rites, legal concerns, insurance, coroners, funeral directors, cemeteries, memorials (gravemarkers), cremation, organ donation and transplantation, sympathy calls, condolence letters, widows and widowers, suicide, and death education.

Huber, L. V., et al. *The Cemeteries*. Volume III of *New Orleans Architecture Series*, Mary Louise Christovich, ed. Gretna, LA: Pelican Pub. Co.; 1974.
Thoroughly illustrated with photographs and drawings, this text highlights New Orleans funerary architecture. It covers cemetery masonry, ironwork, preservation, and history. Includes an appendix of cemetery locations, a selected bibliography, and index. Aboveground burial detailed extensively.

Jackson, Kenneth T. *Silent Cities: The Evolution of the American Cemetery*. New York: Princeton Architectural Press; 1989.

Jordan, Terry G. *Texas Graveyards: A Cultural Legacy*. Austin: University of Texas Press; 1982.

Kull, Andrew. *New England Cemeteries*. Brattleboro, VT: Stephen Greene Press.
A guide to 262 New England cemeteries with information on old cemetery art. Includes detailed maps.

Laas, William. *Monuments in Your History*. New York: Popular Library; 1972.

Lindley, Kenneth. *Of Graves and Epitaphs*. London: Hutchison; 1965.

The Price of Death: A Survey Method and Consumer Guide for Funerals, Cemeteries, and Grave Markers. Washington, DC: US Government Printing Office; 1975.
Note: Consumer Survey Handbook 3. A Federal Trade Commission Publication, Seattle Regional Office.

Schafer, Louis S. *Best of Gravestone Humor*. New York: Sterling; 1990.
Note: Illustrated.

Sloane, David C. *The Last Great Necessity: Cemeteries in American History*. Baltimore: Johns Hopkins University Press; 1991.

Stranix, E. L. *The Cemetery: An Outdoor Classroom*. Philadelphia: Con-Stran Productions; 1977.
Note: A Student Workbook, Project Kare edition.

Tashjian, Dickran. *Memorials for Children of Change: The Art of Early New England Stonecarving*. Middletown, CT: Wesleyan University Press; 1974.

Walker, G. A. *Gatherings from Graveyards*. New York: Arno Press; 1930.
Note: Reprinted in 1977.

Wallis, Charles. *Stories on Stone: A Book of American Epitaphs*. New York: Oxford University Press; 1954.

History of Mortuary Science

Basevi, W. H. F. *The Burial of the Dead*. London: George Routledge and Sons; 1920.
A detailed historical, cross-cultural study of burial and cremation from prehistoric times to the twentieth century.

Bendann, Effie. *Death Customs: An Analytical Study of Burial Rites*. New York: Alfred A. Knopf; 1930.
Note: Also published in 1930 by Kegan, Paul and Co., London.
An examination of the funerary practices of many nations and religious groups from early times to the date of publication. Discusses the relationship between funerary practices and the belief and thought forms of a people. The text is divided into two sections: similarities and

differences (of funeral rites and ceremonies). She covers disposal of the dead, general attitudes toward the corpse, purification, life after death, taboos, mourning, women's connection with funeral rites, totemic conceptions, destruction of property, and cult of the dead. Includes a comprehensive index and glossary of terms.

Bowman, Leroy. *The American Funeral: A Study in Guilt, Extravagance, and Sublimity*. Washington, DC: Public Affairs Press; 1959.
Note: Introduction by Harry A. Overstreet. Also published in 1964 by Paperback Library, New York, and reprinted by Greenwood Press in 1973 and 1975.
The first of the contemporary critiques of the funeral written from the viewpoint of a social scientist who sees the funeral as an anachronism in urban society. He advocates using cremation as a means of making a funeral more economical. Discusses the differentiation between the terms "funeral director," "undertaker," and "mortician." Bowman covers group behavior at funerals, behind-the-scenes activities, family contact with the undertaker, the undertaker's role in the community, and trends in the form and function of funerals.

Clarke, Joseph H. *Reminiscences of Early Embalming*. New York: The Sunnyside; 1917.
An important reference work on the rise of mortuary science education in America.

Dincauze, Dena. *Cremation Cemeteries in Eastern Massachusetts*. Cambridge, MA: Peabody Museum; 1968.

Douglass, Sam P. *Funeral Homes: Legal and Business Problems*. In *Commercial Law and Practice Course Handbook Series*, edited by Roger A. Needham, "A4-1052"–Number 5. New York: Practising Law Institute; 1971.
For annotation, see page 68.

Dowd, Quincy L. *Funeral Management and Costs*. Chicago: University of Chicago Press; 1921.

Note: Subtitled *A World Survey of Burial and Cremation.* An early social scientific study of the funeral. Discusses modern development of cremation and witnessing a cremation.

Eckels College of Mortuary Science Inc. *Modern Mortuary Science,* 4th ed. Philadelphia: Westbrook Publishing Co.; 1958.

Farrell, James J. *Inventing the American Way of Death, 1830-1920.* In *American Civilization Series,* Allen F. Davis. Philadelphia: Temple University Press; 1980.
Farrell contends that death is a cultural event and that societies reveal themselves in their treatment of death. Significant sections include the development of the modern cemetery, the modernization of funeral service, and the cosmological contexts of death. Farrell describes and analyzes the development of the American way of death. It is not like Jessica Mitford's *The American Way of Death,* which focuses on the funeral industry's profit motive. This text emphasizes the complexity of cultural change.

Gale, Frederick. *Mortuary Science.* Springfield, IL: Charles C. Thomas; 1960.

A complete textbook on embalming and restorative art with important references to the history and development of funeral directing and mortuary science. Illustratedwith drawings and photographs.

Goldstein, Lynne G. *Mississippian Mortuary Practices.* In *Scientific Paper Series,* No. 4. Kampsville, IL: Center for American Archaeology; 1980.
Note: Illustrated.

Gorer, Geoffrey. *Death, Grief, and Mourning.* In *The Literature of Death and Dying Series.* New York: Doubleday; 1965.
Note: Published in 1977 by Arno Press, New York.
Though Gorer's text does cover the practicalities of death,

funerals, and their aftermath, it is written primarily from from anthropological and psychological viewpoints. Gorer discusses bereavement extensively, with one section devoted to types of bereavement: death of father, death of child, etc. He also covers telling children about death, the afterlife, issues surrounding the clergy and the church, body disposal (the funeral), family gatherings, gravestones, condolences, and mourning. The four appendixes are: current and recent theories of mourning and the present material, a questionnaire with statistical tables, religious beliefs and practices: 1963 and 1950 compared, and the pornography of death. Also includes index of informants quoted more than once.

Habenstein, Robert W., and William M. Lamers. *The History of American Funeral Directing*. New York: Omnigraphics Inc.; 1990.
Note: Reprint of the original 1955 edition. Second printing in 1956.

For annotation, see page 10.

Huntington, R. and P. Metcalf. *Celebrations of Death: The Anthropology of Mortuary Ritual*. New York: Cambridge University Press; 1979.
Note: Reprinted in 1980, 1981, and 1984. Also cited as *Celebration of Death*. Illustrated.
Particularly important is the conclusion entitled "American Deathways." It includes critiques of the funeral industry.

Johnson, E. C., and G. R. *Alone in His Glory* [unpublished manuscript on Civil War mortuary practices].

Johnson, Edward. *A History of the Art and Science of Embalming*. New York: Casket and Sunnyside; 1944.

Kastenbaum, Robert, ed. *Death and Dying*. New York: Arno Press; 1977.
Note: This title constitutes a 40-volume set. It details

early mortuary practices and documents mortuary science topics through the 20th century.

Martin, Edward. *Psychology of Funeral Service*, 6th ed. Grand Junction, CO: Edward A. Martin.
Note: Third edition published in 1950.
Martin begins the text with a prologue on the importance of education in general and the necessity of mortuary education to society. He discusses a variety of aspects of funeral service from historical background to modern-day practical considerations including an introduction to psychology. He also includes sections on emotion, learning and memory, adjustment to mental conflict, grief, sentiment, religion (with an encyclopedic coverage of 11 religions of the world and 20 religious concepts), funeral rituals (burial, cremation, mutilation, dismemberment, cannibalism, abandonment, and exposure), public relations, and embalming, and a chapter on "psychology in action." Includes the Funeral Service Oath, index, and glossary.

Mitford, Jessica. *The American Way of Death*. New York: Simon and Schuster; 1963.

Note: Also published in 1963 by Fawcett Publications, Greenwich, CT. Numerous reprints.

For annotation, see page 12.

National Funeral Directors Association, 21st Century Committee. *Tradition in Transition*. Milwaukee: National Funeral Directors Association; 1981.

Passing: The Vision of Death in America. Westport, CT: Greenwood Press; 1977.
Contains a history of mortuary customs.

Phipps, William E. *Cremation Concerns*. Springfield, IL: Charles C. Thomas; 1989.
This book presents a balanced analysis of the issues

surrounding cremation. Topics covered include: a history of cremation (ancient pyres), religious opposition, scientific influences, reasons for renewing the practice, a contemporary outlook, Christian acceptance, memorializing options, and pre-planning advantages. Includes extensive notes, a sample Cremation Planning Form, and a subject index. Illustrated.

Pine, Vanderlyn R. *Caretaker of the Dead: The American Funeral Director*. New York: Irvington; 1975.
This text is a thorough discussion of the American funeral service professional. It gives a historical portrait of the profession and discusses professionalism, organization within funeral homes, behavior outside the funeral home, public behavior in the funeral home, non-public behavior in the funeral home, the presentation of self, the funeral director's role, caretakers, and personal service. Appendixes include a funeral directing questionnaire, funeral arranger interview schedule, and findings of a professional census. In his introduction, Pine discusses the study of death.

Polson, Cyril J., R. P. Brittain, and T. K. Marshall. *Disposal of the Dead*. New York: Philosophical Library; 1953.
Note: Also published in 1962 by English Universities Press, London, and by Charles C. Thomas Publishers, Springfield, IL.
A comprehensive study of burial and cremation practices, focusing somewhat on those of England. Contains a thorough historical introduction to the disposal of the dead. Also includes sections on mediate disposal (death certificates, coroners, registration, etc.), cremation, burial (churchyards, cemeteries, burial grounds), funeral rites, exhumation, embalming, and funeral direction. Distinguishes mummification and embalming as modes of preservation. Treats unusual subjects such as preservation of human heads, ship-burial, and radioactive corpses.

Raether, Howard C. *Funeral Service: A Historical Perspective.* Milwaukee: National Funeral Directors Association; 1990.

Rech, Edward H. *Glimpses into Funeral History* [unpublished manuscript]. Cincinnati: produced under the auspices of Hess and Eisenhardt Co.

Rush, Alfred C. *Death and Burial in Christian Antiquity.* Washington, DC: Catholic University of America Press; 1941.

Sloane, David C. *The Last Great Necessity: Cemeteries in American History.* Baltimore: Johns Hopkins University Press; 1991.

Sourcebook on Death and Dying, 1st ed. Chicago: Marquis Professional Publications; 1982.

Stannard, David E., ed. *Death in America.* Philadelphia: University of Pennsylvania Press; 1975.
The author, assistant professor of American studies at Yale University, has written and collected essays on attitudes toward death as a dimension of American culture. The contributors are anthropologists, cultural historians, art historians, and literary scholars. Especially pertinent to this work is Stanley French's "The Cemetery as Cultural Institution."

Liturgies and Sermons

Allen, R. Earl. *Funeral Source Book.* In *Preaching Helps Series.* Grand Rapids: Baker Books; 1984.

Bachmann, C. Charles. *Ministering to the Grief-Sufferer.* Philadelphia: Fortress Press; 1967.
Focuses on pastoral care of the bereaved, but also includes sections on the ministry of a funeral.

Baerwald, Reuben C., ed. *Hope in Grief.* St. Louis: Concordia; 1966.
Offers suggestions for making the funeral a service of worship and for developing the funeral sermon. Also contains a collection of sermons and resources.

Barton, F. M. *One Thousand Thoughts for Memorial Addresses*. New York: Doran.

Bedwell, B. L. *Sermons for Funeral Occasions*. Firm Foundation Pub.; 1960.

Biddle, Perry H. *Abingdon Funeral Manual.* Nashville: Abingdon Press; 1976.
Note: Revised edition. Original edition published in 1976. This text, intended for Christian ministers, contains suggestions for sermons, music, and prayers at funerals. The author includes special material for use at the funeral of a child, for suicide, and for other tragic deaths. Includes detailed liturgies, information on how to develop a church policy on funerals, and instructions on how to conduct funeral services for a variety of Protestant denominations.

Blackwood, Andrew Watterson. *The Funeral: A Source Book for Ministers.* Philadelphia: Westminster Press; 1942.
Note: Also cited as Westminister Press.

Blair, Robert. *The Minister's Funeral Handbook: A Complete Guide to Professional and Compassionate Leadership*. Grand Rapids: Baker Books; 1990.

Cadenhead, Al, Jr. *The Minister's Manual for Funerals.* Nashville: Broadman; 1988.
Cadenhead, a Baptist minister, offers an extensive array of homiletical material, suggestions for pastoral care in the funeral setting, and suggested orders of service and scripture readings with a collection of appropriate poems. He also includes a collection of prayers and benedictions as well as suggestions for further readings.

Champlin, Joseph M. *Through Death to Life: Preparing to Celebrate the Funeral Mass*, rev. ed. Notre Dame, IN: Ave Maria; 1990.

D'Alembert, Jean. *Eulogies.* New York: Gordon Press.

Doyle, Charles H. *Fifty Funeral Homilies*. Christian Classics; 1984.

Ford, Josephine M. *The Silver Lining: Personalized Scriptural Wake Services*. Mystic, CT: Twenty-third; 1987.

Ford, W. Herschel. *Simple Sermons for Funeral Services*. Grand Rapids, MI, 1985.

Funeral Liturgy Planning Guide. Collegeville, MN: Liturgical Press; 1984.

Harmon, N. B. *The Pastor's Ideal Funeral Manual.* Nashville: Abingdon-Cokesbury Press; 1942.

Hutton, Samuel W. *Minister's Funeral Manual.* Grand Rapids: Baker Books; 1968.

International Commission on English in the Liturgy. *Order of Christian Funerals: The Roman Ritual.*

Irion, Paul E. *A Manual and Guide for Those Who Conduct a Humanist Funeral Service.* Baltimore: Waverly Press; 1971.

Keiningham, C. W. *Sermon Outlines for Funerals.* In *Sermon Outline Series*, No. 2. Grand Rapids: Baker Books; 1988.

Leach, W. H. *The Cokesbury Funeral Manual.* Nashville: Cokesbury Press; 1932.
Note: Also cited as *The Cokeburn Funeral Manual*.

Leach, W. H. *The Improved Funeral Manual.* Grand Rapids: Baker Book House; 1956.

The Lord Is My Shepherd-A Book of Wake Services. Notre Dame, IN: Ave Maria Press; 1971.
Contains options to the official text for the funeral liturgy for those planning the wake and funeral Mass.

Marchal, Michael. *Parish Funerals: A Guide to the Order of Christian Church.* Chigago: Liturgy Tr. Publications; 1987.

Meyer, F. B., et al. *Funeral Sermons and Outlines.* Grand Rapids: Baker Books; 1984.
Note: Part of the Pulpit Library.

Motter, Alton M., ed. *Preaching About Death.* Philadelphia: Fortress Press; 1975.

Office of Worship for the Presbyterian Church (U.S.A.) Staff and Cumberland Presbyterian Church Staff. *The Funeral: A Service of Witness to the Resurrection.* In *Supplemental Liturgical Resource Series,* No. 4. Westminster John Knox; 1986.

Order of Christian Funerals: General Introduction and Pastoral Notes. In *Liturgy Documentary Series,* No. 8. Washington, DC: US Catholic; 1989.

Poovey, W. A., ed. *Planning a Christian Funeral: A Minister's Guide.* Minneapolis: Augsburg; 1978.
A book of popular funeral sermons with an introductory text on the purpose of a funeral and the facets of the funeral and burial ceremonies. Biblical text accompanies each sermon, and at the end of each the author lists the preacher, the occasion, and comments.

Rite of Funerals. Washington, DC: United States Catholic Conference; 1971.
This is the official text for the funeral liturgy.

Rutherford, Richard. *Death of a Christian: The Rite of Funerals,* rev. ed. Pueblo, CO: Pueblo Pub. Co.; 1990.

Note: Part of Studies in the Reformed Rites of the Catholic Church: Vol. 7. Originally published in 1980.

A Service of Death and Resurrection. Nashville: Abingdon Press; 1979.
Note: Part of Supplemental Worship Resources, No. 7. An aid for understanding the church's ministry at the time of death and for planning and conducting the funeral service.

St. Gregory Nazianzen and St. Ambrose. *Funeral Orations*. Vol. 22 of *Fathers of the Church Series*. Washington, DC: Catholic University Press; 1953.

Wagner, Johannes, ed. *Reforming the Rites of Death*. Vol. 32 of the *Concilium Series*. Mahwah, NJ: Paulist Press; 1968.

Wallis, Charles Langworthy. *The Funeral Encyclopedia: A Source Book*. Grand Rapids: Baker Book House; 1973.
Note: Originally published in 1953 by Harper and Brothers, New York.
Contains a variety of funeral sermons, poems, prayers, and guidance for the pastor. Text is divided into five sections: the funeral service, a treasury of sermons, an anthology of funeral poems, a sheaf of funeral prayers, and professional conduct. Includes poetry, textual, classification, and topical indexes. All material is attributed to its respective author.

Wood, Charles R., ed. *Sermon Outlines for Funeral Services*. Grand Rapids: Kregel; 1970.

Memorial Societies

Burgess, Vicki. *The Memorial Societies Movement: A Challenge to the Funeral Industry in the USA*. University Park, PA: Pennsylvania Sociological Society; 1979.

Continental Association of Funeral and Memorial Societies. *Bibliography of Death Education*. Washington, DC: Continental Association of Funeral and Memorial Societies.
Note: The Continental Association serves as a clearinghouse for information about the nation's memorial societies.

Continental Association of Funeral and Memorial Societies. *Bibliography of Funeral Reform*. Washington, DC: Continental Association of Funeral and Memorial Societies.

Continental Association of Funeral and Memorial Societies. *Funeral and Memorial Societies*. Washington, DC: Continental Association of Funeral and Memorial Societies; 1974.
Answers questions most frequently asked about funeral and memorial societies.

Continental Association of Funeral and Memorial Societies. *Handbook for Memorial Societies*. Washington, DC: Continental Association of Funeral and Memorial Societies.

Irion, Paul E. *The Funeral: Vestige or Value?* In *The Literature of Death and Dying Series*. New York: Arno Press; 1966.
Note: Also published with Abingdon Press, Nashville, in the same year. Reprinted by Arno Press in 1976.
Based on religious, cultural, social, and psychological understanding of the nature of the funeral. Contemporary practices are evaluated in light of the valuable functions of the funeral, and new designs are

proposed to conserve significant values. Contains sections on cremation and memorial societies.

Mitford, Jessica. *The American Way of Death*. New York: Simon and Schuster; 1963.

Note: Also published in 1963 by Fawcett Publications, Greenwich, CT. Numerous reprints.

For annotation, see page 12.

Myers, James, Jr. *Cooperative Funeral Associations*. New York: Cooperative League of the U.S.A.; 1946.
Note: Pamphlet 409. Author also cited as James Meyers, Jr. Also cited as published in Chicago.
An early publication indicating some reasons for high-priced funerals and offering suggestions on how consumers might cut costs.

Nora, Fred. *Memorial Associations: What They Are--How They Are Organized*. Chicago: Cooperative League of the USA; 1962.
Discusses how volunteers can organize societies to obtain respectable funerals at decent costs.

Mortuary Science Education

Gale, Frederick. *Mortuary Science*. Springfield, IL: Charles C. Thomas; 1960.

Presents a historical perspective on the teaching of embalming and restorative art.

Habenstein, Robert W., and William M. Lamers. *Funeral Customs the World Over*. Milwaukee: Bulfin Printing; 1963.
Note: Reprinted in 1974. First printing in 1960.
The authors present a detailed account of mortuary practices throughout the world. Sections are Asia, the Middle East, Africa, Oceania, Europe, Latin America,

Canada, and the United States. Especially comprehensive are the chapters on Native American funeral customs and burial rites. Illustrated with photographs. Covers all aspects of funeral service: embalming, burial, funeral coaches and automobiles, funeral homes, mortuary science education, etc. Also includes details on Jewish, Latter-Day Saints, and American Gypsy funeral rites.

Habenstein, Robert W., and William M. Lamers. *The History of American Funeral Directing*. New York: Omnigraphics Inc.; 1990.
Note: Reprint of the original 1955 edition. Second printing in 1956.

For annotation, see page 10.

Krieger, Wilber M. *Successful Funeral Service Management*. Englewood Cliffs, NJ: Prentice-Hall; 1951.
This text is written for both funeral home management personnel and potential funeral professionals. It applies general management principles and concepts to the funeral service business. Krieger discusses how to enter the profession (license requirements, education, personal characteristics, etc.), public attitudes toward funeral service, management responsibilities, selecting a location, setting up the organization, financing, required investments (with furniture, fixtures, and equipment checklists), working capital, attracting business through advertising and other means, merchandising, accounting, forms to use, credits and collections, letter writing, employment policies, personnel relations, and ethics. Includes an appendix of state licensing rules for embalmers and funeral directors compiled by O. J. Willoughby, publisher of *Southern Funeral Director*.

Margolis, Otto S., et al., eds. *Thanatology Course Outlines*. New York: MSS Information Co.; 1978.

Martin, Edward. *Psychology of Funeral Service*, 6th ed. Grand Junction, CO: Edward A. Martin.

Note: Third edition published in 1950.
Martin begins the text with a prologue on the importance of education in general and the necessity of mortuary education to society. He discusses a variety of aspects of funeral service from historical background to modern-day practical considerations including an introduction to psychology. He also includes sections on emotion, learning and memory, adjustment to mental conflict, grief, sentiment, religion (with an encyclopedic coverage of 11 religions of the world and 20 religious concepts), funeral rituals (burial, cremation, mutilation, dismemberment, cannibalism, abandonment, and exposure), public relations, and embalming, and a chapter on "psychology in action." Includes the Funeral Service Oath, index, and glossary.

Mayer, Robert G. *Embalming: History, Theory, and Practice*. Norwalk, CT: Appleton and Lange; 1990.

For annotation, see page 2.

Mitford, Jessica. *The American Way of Death*. New York: Simon and Schuster; 1963.

Note: Also published in 1963 by Fawcett Publications, Greenwich, CT. Numerous reprints.

For annotation, see page 12.

Myers, John. *Manual of Funeral Procedure*. Casper, WY: Prairie Publishing Co.; 1956.

National Funeral Directors Association, 21st Century Committee. *Tradition in Transition*. Milwaukee: National Funeral Directors Association; 1981.

Rudman, Jack. *Funeral Directing Investigator*. In *Career Examination Series*, C-3112. Syosset, NY: National Learning; 1988.

Rudman, Jack. *Mortuary Caretaker*. In *Career Examination Series*, C-500. Syosset, NY: National Learning; 1989.

Rudman, Jack. *Mortuary Technician*. In *Career Examination Series*, C-514. Syosset, NY: National Learning; 1989.

Rudman, Jack. *Senior Mortuary Caretaker*. In *Career Examination Series*, C-721. Syosset, NY: National Learning; 1989.

Sourcebook on Death and Dying, 1st ed. Chicago: Marquis Professional Publications; 1982.

Ward, A. E., ed. *Study Courses in Funeral Management*. London: National Association of Funeral Directors; 1948.

Wolfelt, Alan D. *Interpersonal Skills Training: A Handbook for Funeral Service Staffs*. Muncie, IN: Accelerated Development Inc.; 1990.
Note: Distributed by the National Funeral Directors Association, Milwaukee. Also cited as *Interpersonal Skills Training: A Handbook for Funeral Home Staffs*.
Wolfelt, a clinical thanatologist and director of the Center for Loss and Life Transition in Fort Collins, CO, presents a how-to text on counseling, grief therapy, and general interpersonal skills training for funeral service personnel. He addresses the need for such training, noting that mortuary science schools focus primarily on anatomy and physiology and little on counseling, a much needed skill for funeral directors, according to Wolfelt. He extensively covers grief, communication techniques, and mourning. Includes an important section on funeral service stress. Also offers list of training opportunities for funeral directors and an index.

Occupational Health and Safety

Block, S. S. *Disinfection, Sterilization, and Preservation*, 3rd ed. Philadelphia: Lea and Febiger; 1983.

Dorn, James M., and Barbara M. Hopkins. *Thanatochemistry: A Survey of General, Organic, and Biochemistry for Funeral Service Professionals*. Reston, VA: Reston Publishing Co.; 1985.
Note: The authors are funeral service educators at the Cincinnati College of Mortuary Science.
Contains extensive sections on general chemistry, organic chemistry, and biochemistry. Discusses a variety of problems and concerns relating to the embalming process. They include: decomposition, denaturation, enzyme activity, and rigor mortis. Appendixes contain sections on radiation chemistry and a summary of the action and composition of embalming fluids. A textbook approach with chapter summaries and questions.

Eckels, Howard S. *Practical Embalmer*. Philadelphia: H. S. Eckels Co. Publishers; 1903.

Eckels, Howard S. *Sanitary Science: A Reference and Textbook for the Communicable Diseases, Disinfection and Chemistry for the Undertaker*. Philadelphia: G. F. Lasher; 1906.
Note: Illustrated.

Habenstein, Robert W., and William M. Lamers. *The History of American Funeral Directing*. New York: Omnigraphics Inc.; 1990.
Note: Reprint of the original 1955 edition. Second printing in 1956.

For annotation, see page 10.

Hinson, Maude R. *Final Report on Literature Search on the Infectious Nature of Dead Bodies for the Embalming Chemical Manufacturers Association*. Cambridge, MA: Embalming Chemical Manufacturers Association; 1968.
Note: Author was a medical research librarian in Downers Grove, IL.

Mayer, Robert G. *Embalming: History, Theory, and Practice*. Norwalk, CT: Appleton and Lange; 1990.

For annotation, see page 2.

McCurdy, C. W. *The Embalmer as Sanitarian: Embalming and Embalming Fluids.* Wooster, OH: University of Wooster; 1895.
Note: Doctoral manuscript.

National Funeral Directors Association. *Living with OSHA: Employee Handbook.* Milwaukee: National Funeral Directors Association; 1990.

New Jersey Funeral Service Education Corporation and New Jersey State Funeral Directors Association. *OSHA Compliance in Funeral Service.* 1990.

Spriggs, A. O. *The Art and Science of Embalming.* Springfield, OH: Champion Chemical Co.; 1963.
Spriggs, director of service and research for the Champion Company, calls this text the second edition of *The Textbook on Embalming,* the first edition published in 1933. He discusses the history of embalming, arterial embalming, decomposition and preservation, the blood vascular system, body preparation, raising vessels, injection and drainage, cavity treatment, surface tension, discolorations before and after death, diseases of the blood vessels, heart, respiratory system, digestive tract, liver, and kidneys, diabetes, infectious and contagious diseases, heat prostration, preparation of children's bodies, embalming posted bodies, accidental and violent deaths, poisons, health department duties, cause of death, cosmetics, and general anatomy. Includes a compend of 352 questions and answers. Illustrated.

Organ and Body Donation

Carlson, Lisa. *Caring for Your Own Dead.* Hinesburg, VT: Upper Access Publishers; 1987.
A complete guide for those who wish to handle funeral

arrangements themselves. The text is divided into three parts. Part 1 discusses home funerals, cremation, embalming, burial, body and organ donation, and legal issues. Part 2 lists laws, regulations, and services in each state. Includes names and addresses of state agencies and regulatory boards, crematories, and sites for body donation. The final part is a group of appendixes with details on death certificates, preneed spending, grieving, and consumer information on Federal Trade Commission funeral regulation rules. Also includes glossary of funeral-related terms. Text comes with a sample death certificate from North Dakota.

Editors of Consumer Reports. *Funerals: Consumers' Last Rights.* Mount Vernon, NY: Consumers' Union; 1977.

Note: Subtitled *The Consumers' Union Report on Conventional Funerals and Burial...and Some Alternatives, Including Cremation, Direct Cremation, Direct Burial, and Body Donation.* Also published by Pantheon Books, New York.

For annotation, see page 1.

Grollman, Earl A. *Concerning Death: A Practical Guide for the Living.* Boston: Beacon Press; 1974.
Grollman's book is a guide to dealing with the facts and emotions of death. The text contains 20 individually edited sections on the subject of death and funerals--intended primarily for consumers. Pertinent topics covered are: grief, Protestant, Catholic, and Jewish rites, legal concerns, insurance, coroners, funeral directors, cemeteries, memorials (gravemarkers), cremation, organ donation and transplantation, sympathy calls, condolence letters, widows and widowers, suicide, and death education.

Jordahl, Edna K. *Planning and Paying for Funerals.* St. Paul, MN: Agricultural Extension Service, University of Minnesota; 1967.

Note: Revised edition. Also cited as written by Edna K. Fordahl in 1969.
This booklet discusses necessary arrangements for funerals, how to plan for them, financing strategies, laws that relate to death, and body donation.

Mayer, Robert G. *Embalming: History, Theory, and Practice*. Norwalk, CT: Appleton and Lange; 1990.

For annotation, see page 2.

Mitford, Jessica. *The American Way of Death*. New York: Simon and Schuster; 1963.

Note: Also published in 1963 by Fawcett Publications, Greenwich, CT. Numerous reprints.

For annotation, see page 12.

National Funeral Directors Association. *Organ and Tissue Transplantation and Body Donation: A Compendium of Facts Compiled as an Interprofessional Source Book in Cooperation with the University of Minnesota College of Medical Sciences.* Milwaukee: National Funeral Directors Association; 1970.
This booklet lists the requirements of 106 schools of medicine regarding body donations and transplantation of tissues and organs.

Out of Darkness. Minneapolis: Minnesota Lions' Eye Bank, University of Minnesota.

Premature Burial (Vivisepulture)

Tebb, William (F.R.G.S.), and Col. E. P. Vollum, M.D. *Premature Burial*, 2nd ed. London: Swan Sonnenschein and Co. Ltd.; 1905.

Pre-need Service

Better Business Bureau. *The Pre-Arrangement and Pre-Financing of Funerals.* New York: Better Business Bureau; 1960.
Note: Also published in 1963 as *Facts You Should Know, Questions You Should Ask About the Pre-Arrangement and Pre-Financing of Funerals.*

Brown, Harold W. *How to Sell Cemetery Property Before Need.* Topeka, KS: Harold W. Brown; 1975.

Carlson, Lisa. *Caring for Your Own Dead.* Hinesburg, VT: Upper Access Publishers; 1987.
A complete guide for those who wish to handle funeral arrangements themselves. The text is divided into three parts. Part 1 discusses home funerals, cremation, embalming, burial, body and organ donation, and legal issues. Part 2 lists laws, regulations, and services in each state. Includes names and addresses of state agencies and regulatory boards, crematories, and sites for body donation. The final part is a group of appendixes with details on death certificates, preneed spending, grieving, and consumer information on Federal Trade Commission funeral regulation rules. Also includes glossary of funeral-related terms. Text comes with a sample death certificate from North Dakota.

Coleman, Reverend William L. *It's Your Funeral.* Wheaton, IL: Tyndale House; 1979.
Coleman, a Christian pastor, encourages advance planning of death arrangements and provides a general overview of some available options.

Congress, House Select Committee on Aging. *A Guide to Funeral Planning.* Washington, DC: Government Printing Office; 1984.

Draznin, Y. *How to Prepare for Death: A Practice Guide.* New York: Hawthorn Books; 1976.
Draznin's guide, published during what she terms "a bibliographic torrent" of major proportion of books on death. She contends in the preface that death has become society's prime nonfictional fascination. Draznin's text is indeed a practical guide, covering all of the details necessary in preparing for your own death: disposing of the body, the mortuary rites, costs, wills, insurance, and estate conservation. She also includes a 10-chapter section on coping with a death in the family, detailing what to do in the cases of sudden death, accidental death, suicide, homicide, and other circumstances. The text also includes an appendix of supplementary reading notes.

Johannson, Francia Faust, ed. *The Last Rights: A Look at Funerals.* Mills, MD: Owings/Maryland Center for Public Broadcasting; 1975.
Note: Part of the Consumer Survival Kit.
An assortment of articles dealing with funeral planning, cost information, alternatives, and advice.

Jordahl, Edna K. *Planning and Paying for Funerals.* St. Paul, MN: Agricultural Extension Service, University of Minnesota; 1967.
Note: Revised edition. Also cited as written by Edna K. Fordahl in 1969.
This booklet discusses necessary arrangements for funerals, how to plan for them, financing strategies, laws that relate to death, and body donation.

Jowett, Mary W. *A Guide to Funeral Planning.* Independence, MO: Worship Commission, Reorganized Church of Jesus Christ of Latter Day Saints and Herald Publishing House; 1985.

National Funeral Directors Association. *When a Death Occurs: Needs...Concerns...Decisions.* Milwaukee: National Funeral Directors Association; 1974.

Nelson, Thomas. *It's Your Choice: The Practical Guide to Planning a Funeral.* Washington, DC: American Association of Retired Persons (AARP) and National Retired Teachers Association (NRTA); 1987.
Note: Also published in 1982 as *Your Choice: The Practical Guide to Planning a Funeral* by Scott Foresman and Co., Glenview, IL. The 1982 title is also cited as *It's Your Choice: The Practical Guide to Funeral Planning.*

Phipps, William E. *Cremation Concerns.* Springfield, IL: Charles C. Thomas; 1989.
This book presents a balanced analysis of the issues surrounding cremation. Topics covered include: a history of cremation (ancient pyres), religious opposition, scientific influences, reasons for renewing the practice, a contemporary outlook, Christian acceptance, memorializing options, and pre-planning advantages. Includes extensive notes, a sample Cremation Planning Form, and a subject index. Illustrated.

Raether, Howard C., ed. *The Funeral Director's Practice Management Handbook.* Englewood Cliffs, NJ: Prentice-Hall; 1989.
Essentially a textbook, this title exhaustively examines the subject of funeral home management with contributions from 16 industry leaders, scholars and other experts. Raether divides the text into two parts focusing on personalizing professional funeral service practices and practice management and marketing for profitable funeral services. Subjects covered include: grief facilitation, education opportunities for funeral personnel, dealing with the clergy, funeral pricing, public relations, personnel, merchandising, pre-need plans, upgrading funeral service facilities, legal concerns, and FTC regulations. Topical issues are treated thoroughly with sections on "Women in Funeral Service," "Computerization of Recordkeeping," and "Regulatory Changes." Illustrated.

Riley, Miles O'Brien. *Set Your House in Order*. New York: Doubleday; 1980.

Simpson, Michael A. *The Facts of Death*. Englewood Cliffs, NJ: Spectrum/Prentice-Hall; 1979.
Offers information on how to plan one's funeral and estate and how to avoid unscrupulous funeral directors.

Smith, Curtis A. *Help for the Bereaved: What the Family Should Know*. Chicago: Adams Press; 1972.
Discusses funerals, death certificates, financial affairs, and benefits.

United States Senate. *Preneed Burial Service: Hearings Before the Subcommittee on Frauds and Misrepresentations Affecting the Elderly of Special Committee on Aging, U.S. Senate, 88th Congress, Second Session*. Washington, DC: Government Printing Office; 1964.
Discusses preneed funeral and cemetery purchases for senior citizens. Also reveals and documents fraudulent practices.

Professional Associations

American Blue Book of Funeral Directors. New York: Boylston Publications; 1972.

For annotation, see page 20.

Habenstein, Robert W., and William M. Lamers. *The History of American Funeral Directing*. New York: Omnigraphics Inc.; 1990.
Note: Reprint of the original 1955 edition. Second printing in 1956.

For annotation, see page 10.

National Funeral Directors Association. *Resource Manual*. Milwaukee: National Funeral Directors Association; 1979.

Psychology of the Funeral

Bachmann, C. Charles. *Ministering to the Grief-Sufferer*. Philadelphia: Fortress Press; 1967.
Focuses on pastoral care of the bereaved, but also includes sections on the ministry of a funeral.

Biddle, Perry H. *Abingdon Funeral Manual*. Nashville: Abingdon Press; 1976.
Note: Revised edition. Original edition published in 1976.
This text, intended for Christian ministers, contains suggestions for sermons, music, and prayers at funerals. The author includes special material for use at the funeral of a child, for suicide, and for other tragic deaths. Includes detailed liturgies, information on how to develop a church policy on funerals, and instructions on how to conduct funeral services for a variety of Protestant denominations.

Blair, Robert. *The Minister's Funeral Handbook: A Complete Guide to Professional and Compassionate Leadership*. Grand Rapids: Baker Books; 1990.

Bowman, Leroy. *The American Funeral: A Study in Guilt, Extravagance, and Sublimity*. Washington, DC: Public Affairs Press; 1959.
Note: Introduction by Harry A. Overstreet. Also published in 1964 by Paperback Library, New York, and reprinted by Greenwood Press in 1973 and 1975.
The first of the contemporary critiques of the funeral written from the viewpoint of a social scientist who sees the funeral as an anachronism in urban society. He advocates using cremation as a means of making a

funeral more economical. Discusses the differentiation between the terms "funeral director," "undertaker," and "mortician." Bowman covers group behavior at funerals, behind-the-scenes activities, family contact with the undertaker, the undertaker's role in the community, and trends in the form and function of funerals.

Bullough, Vern L. *The Banal and Costly Funeral*. Yellow Springs, OH: The Humanist Association, Humanist House; 1960. Discusses the laws governing the disposal of the dead and high funeral costs. Also covers mourning as a psychological necessity.

Counts, David and Dorothy, eds. *Coping with the Final Tragedy: Dying and Grieving in Cross Cultural Perspective*. Baywood Publishers; 1991.

Dempsey, D. *The Way We Die: An Investigation of Death and Dying in America Today*. New York: McGraw-Hill; 1975.

Draznin, Y. *How to Prepare for Death: A Practice Guide*. New York: Hawthorn Books; 1976.
Draznin's guide, published during what she terms "a bibliographic torrent" of major proportion of books on death. She contends in the preface that death has become society's prime nonfictional fascination. Draznin's text is indeed a practical guide, covering all of the details necessary in preparing for your own death: disposing of the body, the mortuary rites, costs, wills, insurance, and estate conservation. She also includes a 10-chapter section on coping with a death in the family, detailing what to do in the cases of sudden death, accidental death, suicide, homicide, and other circumstances. The text also includes an appendix of supplementary reading notes.

Gorer, Geoffrey. *Death, Grief, and Mourning*. In *The Literature of Death and Dying Series*. New York: Doubleday; 1965.
Note: Published in 1977 by Arno Press, New York.
Though Gorer's text does cover the practicalities of death, funerals, and their aftermath, it is written primarily from

from anthropological and psychological viewpoints. Gorer discusses bereavement extensively, with one section devoted to types of bereavement: death of father, death of child, etc. He also covers telling children about death, the afterlife, issues surrounding the clergy and the church, body disposal (the funeral), family gatherings, gravestones, condolences, and mourning. The four appendixes are: current and recent theories of mourning and the present material, a questionnaire with statistical tables, religious beliefs and practices: 1963 and 1950 compared, and the pornography of death. Also includes index of informants quoted more than once.

Grollman, Earl A. *Concerning Death: A Practical Guide for the Living*. Boston: Beacon Press; 1974.
Grollman's book is a guide to dealing with the facts and emotions of death. The text contains 20 individually edited sections on the subject of death and funerals--intended primarily for consumers. Pertinent topics covered are: grief, Protestant, Catholic, and Jewish rites, legal concerns, insurance, coroners, funeral directors, cemeteries, memorials (gravemarkers), cremation, organ donation and transplantation, sympathy calls, condolence letters, widows and widowers, suicide, and death education.

Irion, Paul E. *The Funeral: An Experience of Value*. Lancaster, PA: Theological Seminary; 1956.
Note: Also cited as a publication of the National Funeral Directors Association, Milwaukee.

Irion, Paul E. *The Funeral: Vestige or Value?* In *The Literature of Death and Dying Series*. New York: Arno Press; 1966.
Note: Also published with Abingdon Press, Nashville, in the same year. Reprinted by Arno Press in 1976.
Based on religious, cultural, social, and psychological understanding of the nature of the funeral. Contemporary practices are evaluated in light of the valuable functions of the funeral, and new designs are

proposed to conserve significant values. Contains sections on cremation and memorial societies.

Irion, Paul E. *The Funeral and the Mourners*. Nashville: Abingdon Press; 1954.
Note: Author also cited as "Ernest F. Irion" for this title.

Jackson, Edgar N. *For the Living*. New York: Channel Press; 1964.
Gives details on how the funeral assists the bereaved. Discusses psychological and theological implications.

Jackson, Edgar N. *The Significance of the Christian Funeral*. Milwaukee: National Funeral Directors Association; 1966.

Kastenbaum, Robert. *Death, Society, and Human Experience*, 3rd ed. Columbus, OH: Charles E. Merrill; 1986.
Note: Second edition published by Mosby in 1981.

Kubler-Ross, Elisabeth. *Death: The Final Stage of Growth*. In *Human Development Books: A Series in Applied Behavioral Science*, Joseph and Laurie Braga, general editors, University of Miami Medical School. Englewood Cliffs, NJ: Prentice-Hall; 1975.
Note: Also published in 1974 by Spectrum Books, New York.
A psychiatrist and well-known authority on death, Kubler-Ross discusses many areas surrounding and encompassing the subject. Rites and customs of American Indians, Jews, Hindus, and Buddhists are covered. An essay entitled "Funerals: Time for Grief and Growth" by Roy and Jane Nichols is included.

Margolis, Otto S. *Grief and the Meaning of the Funeral*. Edison, NJ: Mss Information Corporation; 1975.
Note: Also cited as published in New York.

Marks, Amy Seidel, and Bobby J. Calder. *Attitudes Toward Death and Funerals*. Evanston, IL: The Center for Marketing Sciences, J. L. Kellogg Graduate School of Management, Northwestern University; 1982.

Note: Place of publication also cited incorrectly in some publications as Evansville, IL.

Martin, Edward. *Psychology of Funeral Service*, 6th ed. Grand Junction, CO: Edward A. Martin.
Note: Third edition published in 1950.
Martin begins the text with a prologue on the importance of education in general and the necessity of mortuary education to society. He discusses a variety of aspects of funeral service from historical background to modern-day practical considerations including an introduction to psychology. He also includes sections on emotion, learning and memory, adjustment to mental conflict, grief, sentiment, religion (with an encyclopedic coverage of 11 religions of the world and 20 religious concepts), funeral rituals (burial, cremation, mutilation, dismemberment, cannibalism, abandonment, and exposure), public relations, and embalming, and a chapter on "psychology in action." Includes the Funeral Service Oath, index, and glossary.

National Funeral Directors Association. *Analysis of Attitudes Toward Funeral Directors*. Milwaukee: National Funeral Directors Association; 1948.

National Funeral Directors Association. *Should I Go to the Funeral? What Do I Say?* Milwaukee: National Funeral Directors Association, 1991.

Osborne, Ernest. *When You Lose a Loved One*. New York: Public Affairs Committee; 1965.
Note: Pamphlet 269.
A discussion of the emotional, social, and financial problems relating to death in the family.

Pine, Vanderlyn R., et al. *Acute Grief and the Funeral*. Springfield, IL: Charles C. Thomas; 1976.
Note: Other contributing editors are A. Kutscher, D. Peretz, R. Slater, R. DeBellis, R. Volk, and D. Cherico.

Raether, Howard C., and Robert C. Slater. *The Funeral: Facing Death as an Experience of Life*. Milwaukee: National Funeral Directors Association; 1974.

Sudnow, David. *Passing On*. Englewood Cliffs, NJ: Prentice-Hall; 1967.
Note: Subtitled *The Social Organization of Dying*. Reprinted in 1969.

Restorative Art

Adair, Maude Adams. *The Techniques of Restorative Art*. Dubuque, IA: W. C. Brown Co.; 1948.

Block, S. S. *Disinfection, Sterilization, and Preservation*, 3rd ed. Philadelphia: Lea and Febiger; 1983.

Dhonau, C. O. *Manual of Case Analysis*, 2nd ed. Cincinnati: The Embalming Book Co.; 1928.

Dhonau-Prager. *Restorative Art*. Springfield, OH: Champion Chemical Co.; 1932.

Fredrick, Jerome F. *Embalming Problems Caused by Chemotherapeutic Agents*. Boston: Dodge Institute for Advanced Studies; 1968.

Mayer, J. Sheridan. *Restorative Art*. Philadelphia: Westbrook Publishing Co.; 1943.
Note: Also published by Graphics Arts Press, Livonia, NY, in 1961. Copyright held by Eckels College of Embalming.
Contains detailed instructions on restoring, modeling, and treating all areas of the dead human body. Includes a section on hair restoration. Text also includes a compend of questions at the end of each chapter, bibliography, and index. Illustrated by the author with 566 drawings.

Mayer, Robert G. *Embalming: History, Theory, and Practice.* Norwalk, CT: Appleton and Lange; 1990.

For annotation, see page 2.

Mendelsohn, Simon. *Embalming Fluids: Their Historical Development and Formulation, from the Chemical Aspects of the Scientific Art of Preserving Human Remains.* New York: Chemical Publishing Co.; 1940.
Note: Illustrated.

Myers, E. *Champion Textbook on Embalming.* Springfield, OH: Champion Chemical Co.; 1908.

Papagno, Noella C. *Desairology: The Dressing of the Decedent's Hair*, rev. 2nd ed. In *The Family, the Funeral, and the Hairdresser Series.* Hollywood, FL: J. J. Pub. Co.; 1985.
Note: Author cited on material as Noella Charest-Papagno. First edition published in 1980.
Papagno, having served for over 30 years as a desairologist for funeral homes, offers a practical, how-to guide to dressing the hair of deceased persons, primarily women. She gives a brief introduction to standard funeral practice and then discusses a variety of pertinent topics: sanitation, shampooing techniques, hairsprays, dryers, brushes, and combs, rollers, curl formation, styling and comb-outs, fluff tops and sides, popular styles (including French twists, chignons, and Afros), hair coloring, color sprays and rinses, procedures for problem hair, wigs and other hairpieces, and cosmetics. Text also includes a section on hair structure, growth, and quality. Includes appendix of equipment and supplies. Papagno includes an important section in her introduction on funeral myths that would be helpful to the first-time desairologist.

Papagno, Noella C. *The Hairdresser at the Funeral Home: Desairology Handbook Questions and Answers.* Hollywood, FL: J. J. Pub. Co.; 1981.

Note: Also cited as *The Hairdresser at the Funeral Home--30 Questions and Answers*.

Polson, Cyril J., R. P. Brittain, and T. K. Marshall. *Disposal of the Dead*. New York: Philosophical Library; 1953.
Note: Also published in 1962 by English Universities Press, London, and by Charles C. Thomas Publishers, Springfield, IL.
A comprehensive study of burial and cremation practices, focusing somewhat on those of England. Contains a thorough historical introduction to the disposal of the dead. Also includes sections on mediate disposal (death certificates, coroners, registration, etc.), cremation, burial (churchyards, cemeteries, burial grounds), funeral rites, exhumation, embalming, and funeral direction. Distinguishes mummification and embalming as modes of preservation. Treats unusual subjects such as preservation of human heads, ship-burial, and radioactive corpses.

Renouard, C. A., ed. *Taylor's Art of Embalming*. New York: H. E. Taylor and Co.; 1903.

Selzer, Richard. *Mortal Lessons: Notes on the Art of Surgery*. New York: Simon and Schuster; 1974.
Note: Reprinted in 1975 and 1976.
Contains one section of particular interest to funeral service personnel: "The Corpse." The author, with a tongue-in-cheek approach, makes humorous and sarcastic references to embalming, morgues, autopsies, restorative art, and burial. A summary of the text refers to it as "a rich and stimulating blend of information, reflection, and literary-medical allusion."

Spriggs, A. O. *The Art and Science of Embalming*. Springfield, OH: Champion Chemical Co.; 1963.
Spriggs, director of service and research for the Champion Company, calls this text the second edition of *The Textbook on Embalming*, the first edition published in 1933. He discusses the history of embalming, arterial

embalming, decomposition and preservation, the blood vascular system, body preparation, raising vessels, injection and drainage, cavity treatment, surface tension, discolorations before and after death, diseases of the blood vessels, heart, respiratory system, digestive tract, liver, and kidneys, diabetes, infectious and contagious diseases, heat prostration, preparation of children's bodies, embalming posted bodies, accidental and violent deaths, poisons, health department duties, cause of death, cosmetics, and general anatomy. Includes a compend of 352 questions and answers. Illustrated.

Strub, Clarence G. *The Principles of Restorative Art*.

Thanatology

Bloch, Maurice, and Jonathan Parry, eds. *Death and the Regeneration of Life*. New York: Cambridge University Press; 1982.

Dorn, James M., and Barbara M. Hopkins. *Thanatochemistry: A Survey of General, Organic, and Biochemistry for Funeral Service Professionals*. Reston, VA: Reston Publishing Co.; 1985.
Note: The authors are funeral service educators at the Cincinnati College of Mortuary Science.
Contains extensive sections on general chemistry, organic chemistry, and biochemistry. Discusses a variety of problems and concerns relating to the embalming process. They include: decomposition, denaturation, enzyme activity, and rigor mortis. Appendixes contain sections on radiation chemistry and a summary of the action and composition of embalming fluids. A textbook approach with chapter summaries and questions.

Halporn, R. *The Thanatology Library*. New York: Highly Specialized Promotions; 1976.

Kutscher, Austin H., Jr., and M. A. Kutscher. *A Cross Index of Indices of Books on Thanatology*. New York: MSS Information Co.; 1976.

Kutscher, Austin H., Jr., and M. A. Kutscher. *A Comprehensive Bibliography of the Thanatology Literature*. New York: MSS Information Co.; 1976.

Margolis, Otto S., et al., eds. *Thanatology Course Outlines*. New York: MSS Information Co.; 1978.

Pine, Vanderlyn R. *Caretaker of the Dead: The American Funeral Director*. New York: Irvington; 1975.
This text is a thorough discussion of the American funeral service professional. It gives a historical portrait of the profession and discusses professionalism, organization within funeral homes, behavior outside the funeral home, public behavior in the funeral home, non-public behavior in the funeral home, the presentation of self, the funeral director's role, caretakers, and personal service. Appendixes include a funeral directing questionnaire, funeral arranger interview schedule, and findings of a professional census. In his introduction, Pine discusses the study of death.

Simpson, Michael A., ed. *Dying, Death and Grief: A Critically Annotated Bibliography and Source Book of Thanatology and Terminal Care*. New York: Plenum Press; 1979.

Sourcebook on Death and Dying, 1st ed. Chicago: Marquis Professional Publications; 1982.

When Death Comes. Corvallis, OR: Oregon State University; 1963. Note: Extension Bulletin 809.

Appendix 1: Associations and Organizations

American Board of Funeral Service Education
23 Crestwood Road
Cumberland, ME 04021
The accrediting body for mortuary science and funeral service programs in the United States.

American Cemetery Association
Three Skyline Place, Suite 1111
5201 Leesburg Pike
Falls Church, VA 22041
Owners and managers of cemeteries, suppliers, and professional service firms.

American Institute of Commemorative Art
2446 Sutter Court, N.E.
Grand Rapids, MI 49505
Retailers of memorials and cemetery monuments. Sponsors competitions and awards scholarships.

American Monument Association
6902 North High
Worthington, OH 43085
Manufacturers and wholesalers of granite and marble used in memorials and monuments. Maintains a collection service.

Associated Funeral Directors Service
810 Stratford Avenue
Tampa, FL 33603
Franchise membership granted to one funeral home in a community. Assists funeral directors in shipping human remains.

Casket Manufacturers Association of America
708 Church Street
Evanston, IL 60201

Manufacturers and distributors of burial caskets and other supplies.

Conference of Funeral Service Examining Boards
520 East Van Trees, P.O. Box 497
Washington, IN 47501
National organization of state funeral service examining boards. State boards determine licensing rules and regulations for funeral directors and embalmers.

Continental Association of Funeral and Memorial Societies
2001 S Street, N.W., Suite 530
Washington, DC 20009
Promotes dignity, simplicity and spiritual values in funeral rites and ceremonies. Attempts to reduce unjust funeral costs.

Cremation Association of America
111 East Wacker Drive, Suite 600
Chicago, IL 60601
Offers speakers bureau on cremation and conducts cremation research.

Federated Funeral Directors of America
1622 South MacArthur Boulevard
Springfield, IL 62709
Professional management, accounting and business services for member funeral directors. Compiles statistics on the profession.

Flying Funeral Directors of America
c/o Betty F. Wiley
632 C Avenue
Lawton, OK 73501
Funeral directors interested in aviation. One goal is "to join together in case of mass disaster."

Foundation of Thanatology
630 West 168th Street
New York, NY 10032
Disseminates information regarding the problems inherent in contemplating life-threatening illness and coping with loss,

separation, grief, and bereavement. Based at the Columbia-Presbyterian Medical Center in New York City.

Funeral Home Public Service Group International
1443 East Seventh Street
Charlotte, NC 28204
Encourages community service activity by funeral directors. Seeks to improve public relations for the profession.

Hilgenfeld Foundation for Mortuary Education
P.O. Box 6272
Anaheim, CA 92816
Offers scholarships for research and study to students enrolled in a mortuary science program.

International Cemetery Supply Association
P.O. Box 07779
Columbus, OH 43207
Firms that sell materials, equipment and supplies to the cemetery industry.

International Order of the Golden Rule
Iles Park Place, Suite 315
Springfield, IL 62718
Funeral directors united for public relations, advertising and educational purposes.

Jewish Funeral Directors of America
122 East 42nd Street, Suite 1120
New York, NY 10168
Professional society of Jewish funeral directors.

Monument Builders of North America
1612 Central Street
Evanston, IL 60201
Monument and bronze manufacturers, retailers and wholesalers. Offers special education program.

National Catholic Cemetery Conference
710 North River Road

Des Plaines, IL 60016
Archdiocesan and diocesan directors of Catholic cemeteries. Maintains library on cemeteries, monuments and shrines.

National Concrete Burial Vault Association
P.O. Box 1031
Battle Creek, MI 48016
Manufacturers of concrete burial vaults.

National Foundation of Funeral Service
1614 Central Street
Evanston, IL 60201
An educational foundation maintaining a school of funeral service management extension courses.

National Funeral Directors and Morticians Association
5723 South Indiana Avenue
Chicago, IL 60637
State, district and local funeral directors and embalmers associations and their members. Formerly the National Negro Funeral Directors and Morticians Association.

National Funeral Directors of America
11121 West Oklahoma Avenue
Milwaukee, WI 53227
Federation of state funeral directors associations. The largest professional association of the funeral industry.

National Selected Morticians
1616 Central Street
Evanston, IL 60201
Funeral directors organized to promote consumer information about the profession.

Pre-Arrangement Interment Association of America
1133 15th Street, N.W.
Washington, DC 20005
Cemeteries, funeral homes, and supply companies promoting the pre-planning of funeral arrangements.

Preferred Funeral Directors International
6009 Wayzata Boulevard, Suite 104
Minneapolis, MN 55416
Organized for funeral homes practicing innovative management. Seeks to increase public awareness of the profession and industry.

Telophase Society
1333 Camino Del Rio, South
San Diego, CA 92108
Provides services for cremation and burial at sea.

University Mortuary Science Education Association
P.O. Box 387
Claymont, DE 19703
Professional association for mortuary science educators. Addresses issues in funeral education with literature and professional development activities.

Appendix 2: College and University Programs in Funeral Service Education and/or Mortuary Science

The following programs are accredited by the American Board of Funeral Service Education.

American Academy &
McAllister Institute of
Funeral Service, Inc.
450 West 56th Street
New York, NY 10019
Patrick J. O'Connor,
President
Programs offered: Diploma
(1 yr.), Associate in
Occupational Studies
(312) 757-1190

Bishop State Community
College
Department of Mortuary
Science
351 North Broad Street
Mobile, AL 36603-5898
J.C. Mitchell, Director
Program offered: Associate
in Applied Science (2 yrs.)
(205) 690-6872

Briarwood College
Mortuary Science
Department
2279 Mount Vernon Road
Southington, CT 06489
(candidate for accreditation)

Catonsville Community
College
Mortuary Science Program
800 South Rolling Road
Catonsville, MD 21228
William C. Gonce,
Department
Chairperson
Programs offered:
Certificate (15 mos.),
Associate in Arts
(2 yrs.)
(301) 455-4162

Chicago City Wide College
Department of Mortuary
Science
1900 West Polk Street
Chicago, IL 60612
Louis R. Zefran, Director
Program offered: Associate
in Applied Science (2 yrs.)
(312) 633-5990

Cincinnati College of
Mortuary Science
Cohen Center
3860 Pacific Avenue
Cincinnati, OH 45207-1033
Dan Flory, President
Programs offered: Diploma
(2 yrs.), Associate in
Applied Science (2 yrs.),
Bachelor of Mortuary

Science (3.25 yrs.)
(513) 745-3631

Commonwealth Institute of Funeral Service
415 Barren Springs
Houston, TX 77090
Terry McEnany, President
Programs offered: Diploma (1 yr.), Associate of Science in Funeral Service (5 quarters)
(713) 873-0262

Cypress College
Mortuary Science Department
9200 Valley View Street
Cypress, CA 90630
Douglas G. Metz, Director
Programs offered: Certificate (1 yr.), Associate in Science (2 yrs.)
(714) 826-1131

Dallas Institute of Funeral Service
3909 South Buckner Boulevard
Dallas, TX 75227
Robert P. Kite, President
Programs offered: Diploma (1 yr.), Associate of Applied Science (15 mos.)
(214) 388-5466

Delgado Community College
Department of Funeral Service Education
City Park Campus
615 City Park Avenue
New Orleans, LA 70119-4399
Janice Bodet, Director
Programs offered: Certificate (15 mos.), Associate in Science (2 yrs.)
(504) 483-4014

East Mississippi Community College
Mortuary Science Program
Scooba, MS 39358
Marvin E. Grant, Director
Program offered: Associate of Applied Science (2 yrs.)
(601) 476-8442

Fayetteville Technical Community College
Funeral Service Education Dept.
Post Office Box 35236
Fayetteville, NC 28303
Michael Landon, Chairperson
Program offered: Associate in Applied Science (2 yrs.)
(919) 678-8301

Gupton-Jones College of Funeral Service
280 Mount Zion Road
Atlanta, GA 30354
Daniel E. Buchanan, President

Programs offered:
Diploma (1 yr.), Associate
of Science (2 yrs.)
(404) 761-3118

Hudson Valley
Community College
Mortuary Science
Department
80 Vandenburgh Avenue
Troy, NY 12180
D. Elaine Reinhard,
Chairman
Program offered:
Associate in Science (2 yrs.)
(518) 283-1100

Jefferson State Community
College
Funeral Service Education
Program
2601 Carson Road
Birmingham, AL 35215
James D. Towson, Director
Program offered:
Associate in Applied
Science (2 yrs.)
(205) 853-1200

John A. Gupton College
2507 West End Avenue
Nashville, TN 37203
John A. Gupton, President
Programs offered:
Diploma (1 yr.), Associate
of Arts (2 yrs.)
(615) 327-3927

John Tyler Community
College
Funeral Service Program
Chester, VA 23831
F.W. Thomton, Jr.,
Program Head
Program offered:
Associate in
Applied Science (2 yrs.)
(804) 796-4119

Kansas City Kansas
Community
College
Mortuary Science
Department
7250 State Avenue
Kansas City, KS 66112
David J. Kneib,
Coordinator
Programs offered:
Certificate (16 mos.),
Associate in Applied
Science (2 yrs.)
(913) 334-1100 ext. 226

Lynn University
Mortuary Science Program
3601 North Military Trail
Boca Raton, FL 33431-9990
John A. Chew, Coordinator
Program offered:
Associate in
Science (2 yrs.)
(407) 994-0770

Mercer County
Community College
Funeral Service
Curriculum
1200 Old Trenton Road
P.O. Box B

Trenton, NJ 08690
Robert C. Smith, III, Coordinator
Program offered: Certificate (1 yr.)
(609) 586-4800 ext. 472

Miami-Dade Community College
Department of Funeral Service Education
11380 N.W. 27th Avenue
Miami, FL 33167
Darwin E. Gearhart, Chairman
Program offered: Associate in Science (2 yrs.)
(305) 237-1244

Mid-America College of Funeral Service
3111 Hamburg Pike
Jeffersonville, IN 47130
Glenn A. Morton, President
Programs offered: Diploma (1 yr.), Associate in Applied Science (2 yrs.)
(812) 288-8878

Milwaukee Area Technical College (West Campus)
Funeral Service Department
1200 South 71st Street
West Allis, WI 53214
James Augustine, Coordinator
Program offered: Associate in Applied Science (2 yrs.)
(414) 257-4319

Mount Hood Community College
Department of Funeral Service Education
26000 S.E. Stark Street
Gresham, OR 97030
Garth Nelson, Coordinator
Program offered: Associate in Funeral Service Education (2 yrs.)
(503) 669-6941

Nassau Community College
Mortuary Science Department
c/o S.U.N.Y. College of Technology
Farmingdale, NY 11735
John M. Lieblang, Chairman
Program offered: Associate in Applied Science
(516) 420-2295

New England Institute of Applied Arts and Sciences at Mount Ida College
777 Dedham Street
Newton Center, MA 02159
Louis Misantone, Director

Programs offered:
Diploma (1 yr.), Associate of Science in Funeral Service (2 yrs.)
(617) 969-7000

Northampton Community College
Department of Funeral Service Education
3835 Green Pond Road
Bethlehem, PA 18017
John R. Trout, Director
Programs offered:
Associate in Applied Science (2/3 yrs.), Certificate in Funeral Service Education (2 semesters)
(215) 861-5388

Northwest Mississippi Community College
Funeral Service Technology Program
Desoto Center
Southaven, MS 38671
John M. Mitchell, Chairman
Programs offered:
Certificate (1 yr.), Associate in Applied Science (2 yrs.)
(601) 342-1570

Pittsburgh Institute of Mortuary Science
5808 Baum Boulevard
Pittsburgh, PA 15206
Eugene C. Ogrodnik, President
Programs offered:
Diploma (1 yr.), Associate in Specialized Technology (16 mos.)
(412) 362-8500

St. Louis Community College at Forest Park
Department of Funeral Service Education
5600 Oakland Avenue
St. Louis, MO 63110
Steven B. Koosman, Chairman
Program offered:
Associate in Applied Science (2 yrs.)
(314) 644-9327

San Antonio College
Department of Allied Health Technology
1300 San Pedro Avenue
San Antonio, TX 78284
J. Byron Starr, Chairman
Program offered:
Associate in Applied Sciences in Mortuary Science (2 yrs.)
(512) 733-2905

San Francisco College of Mortuary Science
1363 Divisadero at O'Farrell
San Francisco, CA 94115
Jacquelyn Taylor, President

Programs offered:
Diploma (1 yr.), Associate in Applied Arts (2 yrs.)
(415) 567-0674

Simmons Institute of Funeral Service
1828 South Avenue
Syracuse, NY 13207
Thomas R. Taggart, President
Program offered: Diploma (1 yr.)
(315) 475-5142

Southern Illinois University
Mortuary Science and Funeral Service
Carbondale, IL 62901
George Poston, Director
Program offered: Associate in Applied Science (2 yrs.)
(618) 453-7214

State University of New York
College of Technology at Canton
Mortuary Science Program
Canton, NY 13617
Patrick Fleming, Coordinator
Program offered: Associate in Applied Science (2 yrs.)
(315) 386-7407

University of Central Oklahoma
Department of Funeral Service Education
Edmond, OK 73034-0186
Kenneth Curl, Chairman
Programs offered:
Certificate (2 yrs.),
Bachelor of Science (4 yrs.)
(405) 341-2980

University of the District of Columbia-Van Ness Campus
Mortuary Science Department
4200 Connecticut Avenue, N.W. MB4407
Washington, DC 20008
Leander M. Coles, Chairman
Program offered: Associate in Applied Science (2 yrs.)
(202) 282-7530

University of Minnesota
Program in Mortuary Science
Box 740, UMHC
Harvard & East River Road
Minneapolis, MN 55455
John Kroshus, Director
Program offered: Bachelor of Science (4 yrs.)
(612) 624-6464

Vincennes University

Funeral Service Education
Program
Vincennes, IN 47591
John Alsobrooks,
Chairman
Program offered:
Associate in
Science (2 & 3 yrs.)
(812) 885-4242

Wayne State University
Department of Mortuary
Service
627 West Alexandrine
Detroit, MI 48201
Mary Louise Williams,
Director
Programs offered:
Certificate (3 yrs.),
Bachelor of Science (4 yrs.)
(313) 577-2050

Worsham College of
Mortuary Science
1767 South Wolf Road
Des Plaines, IL 60018
Frederick C. Cappetta,
Chief Administrator
Program offered: Diploma
(1 yr.)
(708) 297-4411

Other schools and programs:

Allen County Community College, Kansas - A
Arrowhead Community College, Minnesota - A
Austin State Junior College, Minnesota - A
Beckley College, West Virginia - A
California College of Mortuary Science - C,A
Central State University, Oklahoma - A
Coffeyville Community Junior College, Kansas - A
Edgecliff College, Ohio - B
Ellsworth Community College, Iowa - A
Florence-Darlington Technical College, South Carolina - A
Gogebic Community College, Michigan - A
Grand Rapids Junior College, Michigan - A
Henry Ford Community College, Michigan - A
Indiana Central University - B
Indiana College of Mortuary Science - A
Kentucky School of Mortuary Science - A
McNeese State University, Louisiana - A
Mesabi Community College, Minnesota - A
Mount Aloysius Junior College, Pennsylvania - A
Northwest Community College, Wyoming - A
Ohio University, Zanesville - A
Point Park College, Pennsylvania - B
Rainy River Community College, Minnesota - A
Rancho Arroyo College of Funeral Service Education, California - A
Southwestern Michigan College - A
S.U.N.Y. Fulton Montgomery Community College, New York - A
S.U.N.Y. Herkimer County Community College, New York - A
University of Wisconsin, Oshkosh - B
Upper Iowa University - B
Wayne State College, Nebraska - B
Wesley College, Delaware - A

A = associate degree.
B = bachelor's degree.
C = certificate.

Appendix 3: Funeral Service State Examining Boards

ALABAMA

Warren S. Higgins, Executive Secretary
State of Alabama
Alabama Board of Funeral Service
Montgomery, AL 36130

ALASKA

Edward R. Mercer, Licensing Examiner
Mortuary Science Section
Department of Commerce and Economic Development
Division of Occupational Licensing
P.O. Box D
Juneau, AK 99811-0800

ARIZONA

Jean A. Ellzey, Executive Director
Arizona State Board of Funeral Directors and Embalmers
1645 West Jefferson, Room 410
Phoenix, AZ 85007

ARKANSAS

John W. Baker, Secretary-Treasurer
Arkansas Funeral Service Board
P.O. Box 2673
Batesville, AR 72503-2673

CALIFORNIA

James B. Allen, Executive Officer
Board of Funeral Directors and Embalmers
1020 N Street, Room 418

Sacramento, CA 95814

COLORADO

Curtis Rostad, Manager
Mortuary Science Commission
Colorado Funeral Directors Association
P.O. Box 1928
Rawlins, WY 82301
(Note: The Mortuary Science Commission was formed in 1982 following the termination of the Colorado State Board of Mortuary Science. There are no requirements for licensure in Colorado, though the commission certifies individuals and firms that meet its rules and requirements.)

CONNECTICUT

Paul E. Driscoll, Chairman
Driscoll Mortuary, Inc.
P.O. Box 299
Torrington, CT 06790

DELAWARE

Gayle Addonizio, Administrative Assistant
Professional Regulation
O'Neil Building
P.O. Box 1401
Dover, DE 19903

DISTRICT OF COLUMBIA

Harriette E. Andrews, Contact Representative
Board of Funeral Directors
Department of Consumer and Regulatory Affairs
Room 923
Occupational and Licensing Administration
614 H Street, N.W.
Washington, DC 20013-7200

FLORIDA

Department of Professional Regulation
Board of Funeral Directors and Embalmers
1940 North Monroe Street, Suite 60
Tallahassee, FL 32399-0750

GEORGIA

Lori Gold, Executive Director
Georgia State Board of Funeral Service
166 Pryor Street, S.W.
Atlanta, GA 30303

HAWAII

Kenneth Fugi, Acting Chief
Sanitation Branch
Department of Health
1250 Punchbowl Street
Honolulu, HI 96813

IDAHO

M.D. Gregersen, Chief
Bureau of Occupational Licenses
2417 Bank Drive, Room 312
Boise, ID 83705-2598

ILLINOIS

Judy Vargas, Professional Services Section Manager
Department of Professional Regulation
320 West Washington Street
Springfield, IL 62786

INDIANA

Mrs, Patricia E. Finney, Board Secretary
Indiana Funeral Service Board

1201 State Office Building
100 North Senate Avenue
Indianapolis, IN 46204

IOWA

Harriett L. Miller, Board Administrator
Professional Licensure
State Department of Health
Lucas State Office Building
Des Moines, IA 50319-0075

KANSAS

Douglas (Mack) Smith, Executive Secretary
Kansas State Board of Mortuary Arts
1200 South Kansas Avenue, Suite 2
Topeka, KS 66612-1331

KENTUCKY

Mary K. Duvall, Executive Secretary
State Board of Embalmers and Funeral Directors
P.O. Box 335
Beaver Dam, KY 42320

LOUISIANA

Lloyd E. Eagan, Secretary
Louisiana State Board of Embalmers and Funeral Directors
P.O. Box 8757
Metairie, LA 70011

MAINE

Bruce Doyle, Regulatory Board Administrator
Licensing and Enforcement
Department of Professional and Financial Regulation
Board of Funeral Service
State House Station #35

Augusta, ME 04333

MARYLAND

Robert C. Adams, Secretary
Maryland State Board of Morticians
4201 Patterson Avenue, 3rd Floor
Baltimore, MD 21215-2299

MASSACHUSETTS

Arthur T. Carow, Executive Secretary
Board of Registration in Embalming and Funeral Directing
Leverett Saltonstall Building
Government Center
100 Cambridge Street
Boston, MA 02202

MICHIGAN

Raymond W. Hood, Sr., Director
Department of Licensing and Regulation
Board of Examiners in Mortuary Science
P.O. Box 30018
Lansing, MI 48908

MINNESOTA

David F. Schwietz, Supervisor
Mortuary Science Unit
717 Delaware Street, S.W.
Minneapolis, MN 55440

MISSISSIPPI

Lindsey Roberts, Jr., Secretary-Treasurer
Executive Office
Mississippi State Board of Funeral Service
802 North State Street, Suite 401
Jackson, MS 39236

MISSOURI

Mrs. Sharlene Rimiller, Executive Director
State Board of Embalmers and Funeral Directors
P.O. Box 423
Jefferson City, MO 65102-0423

MONTANA

Mary Lou Garrett, Administrative Assistant
Board of Morticians
1424 Ninth Avenue
Helena, MT 59620-0407

NEBRASKA

Kris Chiles, Associate Director
Bureau of Embalming Boards
Department of Health
P.O. Box 95007
Lincoln, NE 68509

NEVADA

Edward B. McCaffery, Secretary
Nevada State Board of Funeral Directors and Embalmers
P.O. Box 2462
Reno, NV 89505

NEW HAMPSHIRE

Charles E. Sirc, Secretary
State Board of Registration of Funeral Directors and Embalmers
Health and Human Services Building
6 Hazen Drive
Concord, NH 03301-6527

NEW JERSEY

Maurice W. McQuade, Executive Director
New Jersey State Board of Funeral Service
1100 Raymond Boulevard
Newark, NJ 07102

NEW MEXICO

Geraldine Mascarenas, Administrator
New Mexico State Board of Thanatopractice
725 Saint Michael's Drive
P.O. Box 25101
Santa Fe, NM 87504

NEW YORK

Deborah E. Orecki, Director
Bureau of Funeral Directing
New York State Department of Health
Coming Tower
Empire State Plaza
Albany, NY 12237

NORTH CAROLINA

Corrine Culbreth, Executive Secretary
North Carolina State Board of Mortuary Science
412 North Wilmington Street
Raleigh, NC 27601

NORTH DAKOTA

DeForest Peterson, Executive Secretary
North Dakota State Board of Funeral Service
Box 633
Devils Lake, ND 58301

OHIO

Gordon E. Tatman, Executive Director
The Board of Embalmers and Funeral Directors of Ohio

77 South High Street, 16th Floor
Columbus, OH 43266-0313

OKLAHOMA

Gus Embry, Executive Secretary
State Board of Embalmers and Funeral Directors
4545 North Lincoln Boulevard, Suite 175
Oklahoma City, OK 73105

OREGON

Lucinda J. Potter, Executive Director
Oregon State Mortuary and Cemetery Board
1400 S.W. Fifth Avenue, Room 505
Portland, OR 97201

PENNSYLVANIA

Thomas J. Sturniolo, Administrative Assistant
State Board of Funeral Directors
P.O. Box 2649
Harrisburg, PA 17105-2649

RHODE ISLAND

Robert W. McClanaghan, Administrator
State Department of Health Building
Division of Professional Regulation, Room 104
75 Davis Street
Providence, RI 02908

SOUTH CAROLINA

Avory Bland, Executive Secretary
South Carolina State Board of Funeral Service
P.O. Box 305
Johnston, SC 29832

SOUTH DAKOTA

Floyd A. Miller, Secretary/Treasurer
State Board of Funeral Service
1111 South Main
Aberdeen, SD

TENNESSEE

Bill Teague, Director
Tennessee State Board of Funeral Directors and Embalmers
500 James Robertson Parkway
Nashville, TN 37219

TEXAS

Larry A. Farrow, Executive Director
Texas Funeral Service Commission
8100 Cameron Road, Building B, Suite 550
Austin, TX 78753

UTAH

Kathi Mortensen, Board Secretary
State of Utah
Department of Commerce
Division of Occupational and Professional Licensing
Heber M. Wells Building
160 East 300 South
Salt Lake City, UT 84145-0802

VERMONT

Vermont State Board of Funeral Service
Division of Licensing and Registration
Pavilion Office Building
Montpelier, VT 05602

VIRGINIA

Meredyth P. Partridge, Executive Secretary

Virginia Board of Funeral Directors and Embalmers
Koger Center
1601 Rolling Hills Drive
Richmond, VA 23229-5005

WASHINGTON

Paul Elvig, Program Administrator
Professional Licensing Services
P.O. Box 9012
Olympia, WA 98504-8001

WEST VIRGINIA

John C. McDowell, Executive Secretary
West Virginia Board of Embalmers and Funeral Directors
L & S Building, Fourth Floor
812 Quarter Street
Charleston, WV 25301

WISCONSIN

Department of Regulation and Licensing
Funeral Directors Examining Board
P.O. Box 8935
Madison, WI 53708

WYOMING

David M. Veile, Secretary
Wyoming State Board of Embalming
P.O. Box 349
Worland, WY 82401

ONTARIO, CANADA

Alyson Reynolds, Registrar
Board of Funeral Services
415 Yonge Street, Suite 1609
Toronto, ON, Canada M5B 2E6

Appendix 4: Accreditation Standards of the American Board of Funeral Service Education

Preamble

Funeral Service is a profession practiced by men and women who are required to meet certain educational, societal, and governmental standards. Some are administrative and logistical, while others concern health and sanitation. The primary focus of funeral service is on competent, ethical, service to the public. Accreditation of Funeral Service Educational Programs is intended to help insure that those academic ingredients necessary to the successful practice of funeral service are offered each student in a consistent and universal manner. Standards have been developed to foster this goal.

The Standards of Accreditation as established at diploma, associate degree, and bachelor's degree levels are for the most part qualitative and applied to both individual programs and the institution as a whole. In making its decisions on candidacy or accreditation, the Committee on Accreditation bases its judgment on the objectives of the program, the manner in which it is currently meeting its objective, and the probability that it will continue to meet its objectives in the future. The Standards of Accreditation for the American Board of Funeral Service Education follow.

The minimum Standards of Accreditation are in regular typeface and expressed in terms of "must" and "shall." Guidelines to help in interpreting Standards, where appropriate, are in italics and are expressed in terms of "should" and "may."

(The provisions contained in the Standards are separable. If any section, sub-section, paragraph, sentence, clause, phrase, or requirement contained herein shall be held to be illegal or unenforceable, such illegality or unenforceability of such part shall not affect or in any way impair the validity, application, or enforceability of the remaining portion of this section.)

STANDARD 1: SPONSORSHIP

A. Accreditation is granted to programs within institutions which assume primary responsibility for curriculum content and planning, coordinate classroom teaching and supervise clinical education, appoint faculty, receive and process applications for admission, and grant the diploma or degree documenting completion of the program. Responsibility is also assumed for insuring that clinical experiences assigned each student are educational in nature.

B. The sponsoring institution must be capable of providing for all portions of the required curriculum including all classroom, laboratory, and clinical field experiences specified. It must provide also all service necessary to support the curriculum in funeral service education as specified in the following Standards and must demonstrate evidence of sound financial support of the educational program on a current and continuing basis.

C. The sponsoring institution must be approved by the state in which it is located as evidenced by appropriate state agency approval for the granting of the diploma or degree for which it seeks accreditation.

D. In programs in which more than one institution is involved in the provision of academic and clinical education, responsibilities of the respective institutions including the institution with legal responsibility for program administration, instruction and supervision must be clearly described and documented in a manner signifying agreement by the involved institutions.

Funeral Service Education is essentially a concentration in a professional content area and its support lies within higher education. Programs in this content area are expected to reside within institutions, either single purpose or comprehensive, whose main purpose is education. If associated with, or financially sponsored by, an organization whose main purpose is other than education, care must be taken to insure sufficient separation between the host institution and the funeral service program to foster an effective, independent, and objective learning environment.

STANDARD 2: ORGANIZATION AND ADMINISTRATION

A. The authority and responsibility of each organizational component of the institution (governing board, campus and program administration, faculty, students) together with the processes by which they function or interrelate shall be clearly described by means of a constitution, by-laws or some similar means.

B. The governing board is the legally constituted group which holds the assets of the institution and its objectives and must exercise ultimate program control.

C. The primary task of the administration of the institution and/or program is educational leadership competent to establish conditions providing for good learning opportunities for students, good working conditions for faculty, and good communication processes both inside and outside of the institution/program. A second, but no less important, task of administration is the management of resources in support of educational objectives.

D. A single-purpose institution offering only a program of funeral service education shall be administered by a chief administrative officer, directly accountable only to the governing board for the management of the institution. All other employees shall report directly, or indirectly, to this chief administrative officer. This officer may also be a member of the teaching faculty, as long as the teaching load is reduced commensurate with administrative duties.

E. A funeral service program within a multi-purpose institution shall be administered by a director who has been delegated responsibility for the program. All other employees assigned to the program shall report directly, or indirectly to this person. This officer may also be a member of the teaching faculty, as long as the teaching load is reduced commensurate with administrative duties.

F. The program administrator or institutional chief administrative officer for single-purpose institutions shall possess an academic background consistent with the position of leadership held. For those hired after May, 1979, this will always involve at least the master's degree, within 5 years of the date of the initial full-time appointment in funeral service education, from a regionally accredited college or university or one approved by the U.S. Department of Education.

G. Within program policies, provisions must be made for consideration of student views and judgments in those matters in which students have direct and reasonable interest.

STANDARD 3: AIMS AND PURPOSES

Each program in funeral service education shall have as its central aim recognition of the importance of funeral service personnel as (1) members of a human services profession, (2) members of the community in which they serve, (3) participants in the intimate relationship between bereaved families and those engaged in the funeral service profession, (4) professionals sensitive to and knowledgeable of the responsibility for public health, safety, and welfare in caring for human remains.

Each program shall have at least the following objectives:

1. To enlarge the background and knowledge of students about the funeral service profession.
2. To educate students in every phase of funeral service and to help enable them to develop the proficiency and skills necessary in the profession, as defined in the Preamble above.
3. To educate students concerning the responsibilities of the funeral service profession to the community at large.
4. To emphasize high standards of ethical conduct.
5. To provide a curriculum at the postsecondary level of instruction.
6. To encourage research in the field of funeral service.

STANDARD 4: ADMINISTRATIVE PRACTICES AND ETHICAL STANDARDS

A. Each institution/program must conduct its business and academic activities in an ethical manner. In this regard, each institution/program shall:

1. Publish and adhere to a personnel policy assuring equal employment opportunity for all qualified persons and an admissions policy for students assuring equal consideration for admission without regard to race, color, religion, sex, national origin, age, handicap, marital status, or veteran status.
2. Insure that publications accurately portray the realities of the program/institution; the catalog of each institution must give as much information as possible, including the names and degrees of faculty; the entire curriculum including course names, summary descriptions, and credit hours; requirements for admission, withdrawal, and graduation, and financial policies for tuition, fees, and refunds; all published policies must apply equally to all students.

B. Permanent records must be protected against fire and theft by locked and fire-resistant materials. Permanent records will include the academic transcript of the student, which the institution is obliged to maintain in perpetuity.

C. Policies must exist with regard to due process for both faculty and students, including grievance procedures as well as clearly defined disciplinary policies.

STANDARD 5: FINANCE

A. The institution/program must have financial resources which are adequate to provide instruction and facilities in compliance with this Manual and to ensure graduation of each class accepted.

B. The library budget shall be adequate for proper support of the curriculum.

C. Acceptable accounting practices must be employed as evidenced by current audited financial statements. For institutions which are regionally accredited, institutional compliance will be assumed. In such cases departmental budgets must reflect adequate financial support of the funeral service department. Budget development for the department must be in evidence.

D. The financial accounting system for institutions must not be combined with financial affairs of any other organization.

STANDARD 6: CURRICULUM

A. The program must have a well-organized curriculum plan.

1. The plan must follow a logical sequence and result in a credential appropriate to the length and depth of the curriculum.
2. Semester or quarter credit hours must be assigned to all courses and apportioned appropriately.
3. Course objectives must be established and made known to students
4. All required prerequisite courses shall be clearly indicated.

The curriculum undertaken by funeral service education students generally consists of at least two components: general education course work,and the concentration courses in Funeral Service Education, the so-called "major." The percentage of the curriculum taken by general education courses may vary according to the program in which the student is enrolled. The ABFSE recognizes the associate degree as the minimum recommended educational standard for preparation for the funeral service profession. The specific organization of each curriculum should be based on individual program conditions and situations. The manner of inclusion of ABFSE-stipulated content is left to the campus to decide, as long as the minimum content as stipulated below is included.

B. The concentration in Funeral Service Education shall consist of not less than 50 semester (75 quarter) credits, spread over at least three semesters or four quarters. Following are minimal requirements for the concentration in Funeral Service Education:

1. Public Health and Technical: 14 minimum semester (21 quarter) credits. The curriculum must involve study in the following content areas:
Chemistry, Microbiology and Public Health, Anatomy, Pathology, Embalming, Restorative Art.

Clinical requirements:
Each student shall actively participate in an on-campus course in which the application of Restorative Art principles is practiced in a laboratory setting.

2. Business Management: 14 minimum semester (21 quarter) credits. The curriculum must involve a distribution of study in the following content areas:
Accounting, Funeral Home Management and Merchandising, Computer Application, Funeral Directing, Small Business Management.

3. Social Sciences: 8 minimum semester (12 quarter) credits. The curriculum must involve a distribution of study in the following content areas:
Dynamics of Grief, Counseling, Sociology of Funeral Service, History of Funeral Service, Communication Skills (oral and written).

4. Legal, Ethical, Regulatory: 3 minimum semester (4 quarter) credits. The curriculum must involve study in the following content areas:
Mortuary Law, Business Law, Ethics.

5. Electives: Sufficient to meet graduation requirements for the associate degree or baccalaureate degree as described in each state and institution, if the student is enrolled in a degree program.

C. Instruction shall be at a level generally held commensurate with postsecondary education directed toward the individual growth of each student in areas such as independent thought, resourcefulness, and scientific inquiry.

Appendix 5: Mortuary Science Periodicals

American Cemetery
1501 Broadway
New York, NY 10036
(212) 398-9266

American Funeral Director
Kates-Boylston
Publications, Inc.
1501 Broadway
New York, NY 10036

Canadian Funeral Director
1658 Victoria Park Ave.
Suite 5
Scarborough, ON M1R 1P7
Canada
(416) 755-7050

Canadian Funeral News
Sage Brush Ventures Ltd.
Suite 105
Stockmans Centre
2116 Twenty-Seventh Ave.
NE, (Box 6)
Calgary, AB T2E 7A6
Canada

Casket and Sunnyside
274 Madison Ave.
New York, NY 10016
(212) 685-8310

The Catholic Cemetery
National Catholic
Cemetery Conference
710 North River Rd.
Des Plaines, IL 60016
(312) 824-8131

Cemetery Business & Legal Guide
Clark Boardman Co., Ltd.
435 Hudson St.
New York, NY 10014
(800) 221-9428

Cemetery Management
American Cemetery
Association
Three Skyline Place
Suite 1111
5201 Leesburg Pike
Falls Church, VA 22041

Champion Expanding Encylopedia of Mortuary Practice
Department of Research &
Educational Programs
The Champion Co.
400 Harrison St.
Springfield, OH 45505
(513) 324-5681

Cremationist of North America
Box 7047
Incline Village, NV 89450
(702) 831-3848

The Director
135 West Wells St.
Suite 600
Milwaukee, WI 53203

(414) 276-2500

The Dodge Magazine: Dedicated to Professional Progress in Funeral Service
Dodge Chemical Co.
165 Rindge Ave. Extension
Cambridge, MA 02140
(617) 661-0500

Florida Funeral Director
P.O. Box 6009
Tallahassee, FL 32301
(904) 878-2163

The Forum
New Jersey State Funeral Directors Association, Inc.
3279 Kennedy Blvd.
Jersey City, NJ 07306
(201) 653-0270

Funeral Monitor
3440 John St.
Los Angeles, CA 90026-4513
(800) 453-1199

Funeral Service Insider
ATCOM, Inc.
2315 Broadway
New York, NY 10024-4397
(212) 873-5900

Jewish Funeral Director
12 East 42nd St.
Suite 1120
New York, NY 10168
(212) 370-0024

The Knight Letter
The International Order of the Golden Rule
929 South Second St.
Springfield, IL 62704
(217) 544-7428

Monument Builder News (MB News)
1612 Central St.
Evanston, IL 60201
(312) 869-2031

Morticians of the Southwest
2830 West Kingsley Rd.
Garland, TX 75041
(214) 840-1060

Mortuary Management
1010 Venice Blvd.
Los Angeles, CA 90015
(213) 746-0691

National Funeral Director and Embalmer
734 West 79th St.
Chicago, IL 60620
(312) 487-3600

National Reporter
National Research & Information Center
1614 Central St.
Evanston, IL 60201
(312) 328-6545

The New England Funeral Director
294 Washington St.

Suite 446
Boston, MA 02108
(617) 426-2670

The Oklahoma Funeral Director
Oklahoma Funeral Directors Association
1100 Classen Dr.
Suite 209
Oklahoma City, OK 73103
(405) 236-0561

Red Book
American Monument Association
933 High St.
Suite 220
Worthington OH 43085-4046

Southern Cemetery
John W. Yopp Publications
P.O. Box 7368
Atlanta, GA 30357
(404) 811-9780

The Southern Funeral Director
John W. Yopp Publications
P.O. Box 7368
Atlanta, GA 30357
(404) 811-9780

Stone in America
American Monument Association
6902 North High St.
Worthington, OH 43085
(614) 885-2713

The Tarheel Director
Funeral Information Center
4208 Six Forks Rd.
Suite 357
Raleigh, NC 27609
(919) 781-3549

The Texas Director
Texas Funeral Directors Association, Inc.
1513 South Interstate 35
Austin, TX 78741
(512) 442-2304

Thanatos
P.O. Box 6009
Tallahassee, FL 32301

Title Index

About the Author

JOHN F. SZABO (B.A., University of Alabama; M.I.L.S., University of Michigan) is Director of the Robinson Public Library District in Robinson, Illinois. Formerly Head Librarian at the University of Michigan's Residential College, he has studied the literature of mortuary science and the funeral industry for several years.